Hot Science is a series exploring [...] and technology. With topics from big data to rewilding, dark matter to gene editing, these are books for popular science readers who like to go that little bit deeper ...

T0029144

## AVAILABLE NOW AND COMING SOON:

# EYES

## IN

### Space Telescopes from Hubble to Webb

# THE

# SKY

Andrew May

ICON

Published in the UK and USA in 2024 by
Icon Books Ltd, Omnibus Business Centre,
39–41 North Road, London N7 9DP
email: info@iconbooks.com
www.iconbooks.com

ISBN: 978-183773-127-5
eBook: 978-183773-128-2

Typeset by SJmagic DESIGN SERVICES, India.

Printed and bound in the UK.

## ABOUT THE AUTHOR

Andrew May is a freelance writer and former scientist, with a PhD in astrophysics. He is a frequent contributor to *How It Works* magazine and the Space.com website and has written five other books in Icon's Hot Science series: *Destination Mars, Cosmic Impact, Astrobiology, The Space Business* and *The Science of Music*. He lives in Somerset.

# CONTENTS

CONTENTS

# SPACE AND TELESCOPES 1

Telescopes and space: the two are virtually inseparable. Most of what we know about the universe beyond our own planet is down to telescopes. To say there are craters on the Moon, or that the planet Jupiter – no more than a bright star to the naked eye – is a giant world with moons of its own, would have sounded unbelievable before these facts were revealed by the first telescopes in the 17th century. Yet today, they have been so thoroughly absorbed into our culture that everyone takes them for granted.

The advent of space travel in the 20th century gave us a new perspective on outer space – or at least the nearby part of it represented by our own Solar System. Humans themselves have only ventured as far as the Moon, but robotic probes have travelled much further. Several have visited Jupiter and its moons, including the Juno spacecraft currently in orbit there, while New Horizons is flying through the Kuiper Belt far beyond the orbit of Neptune. Everyone has marvelled at the high-definition images sent back by these missions, such as Juno's panoramic views of

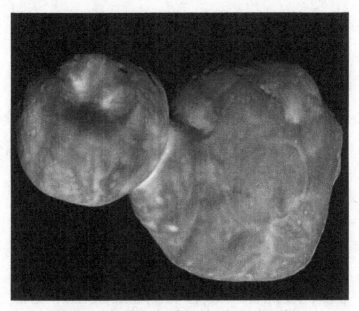

**A composite image of Arrokoth produced by New Horizons using its LORRI (Long Range Reconnaissance Imager) telescope in 2019.**

NASA

Jupiter's swirling, multicoloured clouds and the intriguing glimpses of the 4.5-billion-year-old ice world Arrokoth captured by New Horizons. What's rarely mentioned is the fact that the probes obtained these images using – you guessed it – *telescopes*.

As for the vast universe beyond the Solar System, no human-built spacecraft is going to travel there for a very long time. Yet, if you've ever been blown away by breath-taking photographs of distant nebulae or galaxies – and who hasn't? – they were almost certainly taken by a spacecraft. Not an outward-travelling space probe in this case, but one orbiting our own planet at an altitude of just 540 kilometres. This, of course, is the Hubble Space Telescope – probably the most famous juxtaposition of the words 'space' and 'telescope' of all.

Hubble and its successor the James Webb Space Telescope (JWST) are operated by NASA, the same organisation responsible for the Apollo lunar landings and interplanetary probes like Juno and New Horizons. An American government agency, NASA stands for National Aeronautics and Space Administration – and strictly speaking, the word 'space' in its name refers to space travel. The wider use of the word to encompass the whole of astronomy and cosmology wasn't originally part of the agency's remit – but thanks to Hubble and other space-based telescopes, NASA has become a world authority in those areas too. In fact, the two uses of the word – to mean space travel and the study of the cosmos – have become intertwined. Arrokoth, for example, was discovered by the Hubble telescope in 2014, while it was searching for suitable destinations for the already-launched New Horizons spacecraft.

Yet unlike New Horizons, Hubble has never been anywhere near Arrokoth, which is around 6.6 billion kilometres from Earth. Hubble, at the best of times, is a mere 540 kilometres closer to Arrokoth. And when you turn to the majestic, star-studded galaxies that Hubble is most famous for photographing, the distances become unimaginably big – a trillion times a trillion kilometres or more – so it's not immediately obvious that Hubble has any advantage over an Earthbound telescope at all. Understanding why it does, and why the same is true of JWST and other space-based telescopes like the planet-hunters Kepler and TESS (Transiting Exoplanet Survey Satellite), is one of the aims of this book. The other is to take a closer look at some of the many amazing discoveries made with these telescopes.

But before that, it's worth taking a moment to consider a couple of even more basic questions. Just what is a telescope, and why is it such a useful tool for astronomers?

## Telescope basics

Telescopes have been associated with astronomical observation ever since the time of Galileo Galilei, one of the great pioneers of experimental science, in the early 17th century. But the basic idea of the telescope wasn't his. It started out as a kind of toy called a 'spyglass', made by placing two spectacle lenses at either end of a long tube. Looking through the tube made distant objects appear closer than they really were. In 1608, an enterprising Dutchman named Hans Lipperhey applied for a patent on just such a device, only to have the application turned down on the grounds that the idea was already common knowledge. In fact, similar toys were available for purchase in several European countries by that time.

Lipperhey-style spyglasses only really had novelty value, as opposed to any practical use, since they could only magnify an image by a factor of three or so. The amount of magnification was set by the strength ratio of the two lenses used, and since people were using ready-made spectacle lenses for the purpose, three was about the best they could do. One of Galileo's first innovations was to produce a much stronger, custom-built lens for the eyepiece end of his telescope, allowing him to achieve a much higher magnification.

Galileo's other great breakthrough was to use his telescope to look at astronomical objects that, up to that time, had only ever been seen with the naked eye. He used a telescope with a magnification of 30× to observe the Moon, for example, and described the results in a short treatise called *Sidereus Nuncius* (Latin for 'Starry Messenger') that was published in 1610:

> It is a most beautiful and a very pleasing sight to look at the body of the Moon, which is removed from us by almost 60 terrestrial radii, and to see it as if it were only two radii away. This means that the Moon's diameter looks almost 30 times larger ... Anyone can grasp for himself that the Moon's surface is not smooth and polished but rough and uneven. Like the face of the Earth, it is covered all over with huge bumps, deep holes and chasms.

Only parts of Galileo's 30× telescope survive today, but it's believed that it had a length of around 1.7 metres and a diameter of 40 millimetres at the main lens. These measurements effectively correspond to the two most important parameters of any telescope, its focal length and aperture. The first is the distance at which the main lens (or mirror, in a reflecting telescope) brings incoming light to a sharp focus, while the aperture is simply the diameter of that lens (or mirror). If you see a telescope characterised by just a single dimension, then it's the aperture that's being referred to. So an amateur astronomer boasting, as they're wont to do, of having a 'ten-inch reflector' isn't saying that it's all of ten inches long (about 25 centimetres) but that it has a mirror of that diameter.

When I said a telescope has two important parameters, you may wonder why I didn't mention the one that was so important to Galileo: magnification. But magnification isn't really a property of a telescope *per se*, so much as the eyepiece that's used with it – and that's why we don't need to worry about it in this book. Space telescopes don't have eyepieces, they have sensors similar to the ones used in digital cameras. For that matter, the same is true of virtually all the ground-based telescopes used by professional astronomers these days, and many amateur astronomers too.

In a camera, the equivalent of increasing the magnification is 'zooming in'. A zoom lens is simply one that has a variable focal length, and making this longer causes the object you're looking at to fill a larger portion of the field of view. Inside the camera, the image of the object spans a greater number of sensor pixels at high zoom than it does at low zoom.

This effect is the digital counterpart to the magnification produced by Galileo's eyepiece, and the technical term for it is 'resolution'. In order to explain what this is, and how it differs from traditional magnification, we're going to have to get to grips with one of the more brain-twisting concepts in this book – but one that's absolutely fundamental to the way telescopes are used in astronomy. I'll start by asking a rhetorical question: which is bigger, the Sun or the Moon? I'm sure everyone knows the answer: the Sun is by far the bigger of the two – but it's also further away by roughly the same factor. This means the *apparent* sizes of the Sun and Moon in the sky are pretty much the same (that's why total eclipses work out as neatly as they do). If you imagine looking at the full Moon (I don't want you looking straight at the Sun, even in a thought experiment) and drawing straight lines to opposite edges of it, then the angle between the two lines would be around half a degree. That's what astronomers call the 'angular size' of the Moon – and it's the angular size of the Sun, too.

That's the 'brain-twisting concept' I warned you about: astronomers like to measure objects in terms of angles rather than linear measurements. To complicate matters further, they're usually talking about very small angles, so they divide a degree into 60 arcminutes, and an arcminute into 60 arcseconds, by analogy with minutes and seconds of time. An arcsecond is small, but not unimaginably small – about the same as a five-pence coin (or an American dime) viewed at a distance of four kilometres.

Now let's go back to the question of magnification versus resolution. When Galileo described his 30× magnified view of the Moon, he used words anyone can understand. He said that his telescope made the Moon look 30 times closer than it really is, or 30 times bigger. He didn't need to say anything about angular sizes, even though that's what he was talking about. He meant that when viewed through the eyepiece of his telescope, the Moon appeared to be fifteen degrees across, compared to just half a degree when seen without a telescope. But he didn't need to say that, because the magnification is just the first number divided by the second.

Unfortunately, we can't apply such simple logic to a photograph taken with a digital camera. If you point your smartphone or digital camera at the Moon, set it to maximum zoom and take the best picture you can, what magnification is that? It might look tiny on the device's screen, but what if you take it indoors and look at it on a 50-inch TV screen? What matters here isn't how big the image looks, but how many of the camera sensor's pixels the Moon spans. If the answer is, say, 1,000, that means you've got around half a degree, or 1,800 arcseconds, spanning those thousand pixels. This gives you a resolution of 1.8 arcseconds per pixel – and it turns out that's the most meaningful analogue of magnification for a digital system.

It's the same with telescopes, up to and including Hubble. At this point, if you've been following closely, you may be getting the germ of an idea. 'What if I just buy a super-high-resolution sensor that's millions of pixels across?' you might be thinking. 'Will that make my backyard telescope just as powerful as Hubble?' Unfortunately, it's not that simple. You might be able to get as many pixels across the image as a giant professional telescope, but the result isn't going to be any clearer than it was with a fraction of those pixels. The laws of optics put a natural limit on a telescope's resolution that's inversely proportional to its aperture,* so a small telescope is never going to be as good as a large one. That's one of the reasons that serious astronomical telescopes are designed with the biggest aperture that circumstances allow (the other reason being the obvious one that a large aperture can collect more light than a small one).

The other feature common to virtually all the telescopes used in professional astronomy, besides their enormous size, is the fact that light is collected and brought to a focus by a mirror rather than a lens. At first sight this may seem puzzling. A lens-based telescope – technically called a refractor, from the word 'refraction', describing the bending of light as it passes from one medium to another – was good enough for Galileo, and it's still by far the commonest arrangement in telescopes designed for non-astronomical use. Cameras,

---

* For the technically minded, if a telescope has an aperture of D millimetres, then its theoretical best possible resolution is $134/D$ arcseconds.

too, have lenses at the front rather than mirrors at the back. So why do astronomers do things differently?

The biggest disadvantage of a lens stems from a basic property of refraction called chromatic aberration. This means that as white light passes through a lens, its constituent colours are bent through different angles. This can be a useful effect when we want it to happen, for example, when a prism is used to produce a spectrum of light, but it's extremely irritating when trying to create a sharp image with a lens. Different colours from the same object come to a focus in slightly different places, creating a blurred image. In cameras and small telescopes, the effect can be lessened by using multiple lens elements, but this becomes prohibitively expensive for the larger apertures needed by astronomers.

A better solution is to use a convex mirror to focus the light instead of a lens. Mirrors, of course, work by reflecting light from a surface, rather than refracting it through a material, so they don't suffer from chromatic aberration. A mirror also has the advantage of only needing to be shaped on one side, so it's easier to make. The latter point made reflecting telescopes popular with amateur astronomers back in the days before space became a hot topic, when astronomy was such a niche hobby that you couldn't buy a good-quality, ready-made astronomical telescope in the way you can today. Instead, people had to make their own telescopes, using a simple design invented by Isaac Newton a few decades after Galileo's time.

Despite their different optics, Newtonian reflectors have one very basic thing in common with refracting telescopes: the length of the tube is approximately the same as the telescope's focal length. But if you look up the specs of any large astronomical telescope, whether in space or

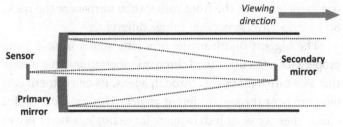

**A simplified diagram illustrating the basic
principle of a Cassegrain reflecting telescope.**

on the ground, you'll find that they generally have a focal
length that's much longer than the physical length. That's
because instead of Newton's design, they use one developed
by Laurent Cassegrain, a French contemporary of Newton.
He came up with a clever way to 'fold' the light rays inside a
telescope tube that results in a more compact and practical
arrangement than Newton's. The downside of Cassegrain
telescopes is that they're more complicated to build, which
is why amateurs traditionally stuck to Newtonians,* but the
Cassegrain has always been the design of choice for professional astronomers.

If the purpose of a telescope was simply to produce an
image of a distant object – which, for most amateur astronomers, that really is all there is to it – then we've already
covered all the basic physics we need to know about the
subject. In professional astronomy, however – the kind that's
done with Hubble and the like – there's much more to using a
telescope than creating pictures. In reality, astronomers don't
spend any more time than the rest of us looking at all those

---

* Although today, thanks to mass production, you can buy a really
good-quality Cassegrain-style telescope for a few hundred pounds.

stunning images the Hubble press office puts out. Those are really just 'outreach' for the general public's benefit.

Science is much more about measurements – putting hard numbers on things – than images. There are some measurements, such as the diameter of a galaxy or the way brightness varies across it, that can be gleaned directly from a digital photograph, but astronomers need to know more than that. If you've ever read an article about a distant galaxy, for example, it may have referred to the speed at which the stars inside it are moving, or what chemical elements they're composed of. You can't get that kind of information by simply looking at a photograph, so where does it come from?

To answer this question, we need to look in a bit more detail at the nature of light. It's a common enough word, and one that I've used multiple times in this chapter already, but what is light, exactly? Confusingly, there are two possible answers, both of which are true even though they sound like they contradict each other. Ultimately, it all depends on what the light is doing. If it's landing on a digital sensor inside a camera or telescope, or the retina of your eye, then you can think of it as a stream of discrete particles called photons, each of which carries a specific amount of energy that the sensor detects. When the light was emitted from its source, whether that was a light-emitting diode (LED) in your room or a star in a distant galaxy, it behaved like a stream of discrete photons too.

Alternatively, when the light is anywhere between the emitter and detector – whether it's travelling through empty space or some other medium like the Earth's atmosphere or a lens – it behaves more like a wave than a stream of particles. Just what it is that's doing the waving is something we'll come back to later, but one thing these waves have in common with photons is the speed they travel at. In empty space, that's around 300 million metres per second – which, needless to say, is very fast. In fact, it's worth taking a moment to look at the implications of this speed, before getting into more detail on the wave nature of light.

If an LED in your room is three metres away, the light from it takes just a hundredth of a millionth of a second to reach your eye, which is so fast as to be almost instantaneous. But it's not literally instantaneous, as we can see if we think about a light source further away, such as the Sun. At a distance of 150 million kilometres, it takes sunlight 500 seconds, or just over eight minutes, to get to us. When we move on to other stars beyond the Sun, the light from them takes years to reach us. They're so far away that it becomes meaningless to talk about their distance from us in kilometres, since most people's minds start to boggle at any figure greater than a trillion. Yet a trillion kilometres doesn't even get you to the nearest star, Proxima Centauri.

It's much easier to say that Proxima is 4.2 light years away, meaning that light from it takes 4.2 years to reach us (or, equivalently, that when we look at it, we see it as it was 4.2 years ago). The light year is one of the most useful concepts in astronomy, and one that we'll encounter many times in this book. It's a great way to talk about astronomical distances without using mind-bogglingly huge numbers. For example, the centre of our own galaxy is around 30,000 light years away, while the distance to Andromeda, our closest neighbouring galaxy, is 2.5 million light years, and the most distant galaxies observed by Hubble are several billion light years distant.

The speed of light also helps to understand two properties associated with the wave theory of light, namely wavelength and frequency. These terms are in common usage in the context of radio waves, but when it comes to light, they're much less familiar than an everyday word that's closely related to them – and that's colour. Take green light, for example, which has a wavelength of around 500 nanometres (500 billionths of a metre). If you picture a wave where the crests are separated by this distance, travelling past you at the speed of light (300 million metres per second), then you'll count 600 trillion crests, or wave cycles, per second.* Since the prefix for trillion is tera, and the technical term for 'cycles per second' is hertz, this means the frequency of a green light wave is 600 terahertz.

Looking at the colour of light from a star or nebula is useful to astronomers because it tells them something about how that light is produced. At this point, we need to briefly switch back to the particle view of light – which, as I said a moment ago, is more relevant to the emission and absorption of light than the wave model. It turns out that what looks like frequency when you think of light as a wave corresponds to the amount of energy per photon in the particle model. The hotter an object is, the higher the energy of the photons it emits – and hence the higher the frequency when the resulting light is viewed as a wave. In terms of colour, blue light has a higher frequency (and shorter wavelength) than red light – so that means, for example, that a blue star is hotter than a red star. You might think that things should be the other way around because the standard convention – for example, on weather maps or bath taps – uses red to mean hot and blue to mean cold. But if you think of metal being

---

* This is a thought experiment, so you're allowed to count superhumanly fast.

heated up in a forge, you can see that the convention is actually wrong: metal glows red first, then changes to blue-white as it gets hotter.

You can see the colour of a star, and hence estimate its temperature, directly from a photograph. But a lot more information can be gained if you break the light down into a spectrum, separating out all the different colours (or frequencies/wavelengths). We already know one way to do this, via the bending of light through a lens, because it's the source of chromatic aberration in refracting telescopes. A prism is a much more effective approach, and the first spectrum-producing instruments were indeed based on the refraction of light through prisms.

Those early instruments involved looking through an eyepiece like a traditional telescope and were called spectroscopes by analogy. The basic process of analysing spectra is still referred to as spectroscopy to this day, although the instruments used – which nowadays project the spectrum onto digital sensors – are called spectrometers or spectrographs (two terms that, as far as I can tell, mean exactly the same thing). Modern ones, such as those on board the Hubble telescope, don't use prisms but a different physical principle called diffraction. But the end result is the same – they split light into a spectrum.

Viewing its spectrum allows astronomers to measure how much energy a star or nebula emits at different wavelengths, but it tells them much more as well. Within the spectrum of an object, a number of sharp lines appear at specific wavelengths that are characteristic of the chemical elements that produced the lines. To understand why, we need to look inside the atoms making up the element. The electrons in each atom can only exist at certain discrete energy levels, meaning that when they jump from one level to another, the exact difference in energy has to be emitted (if they're jumping down) or absorbed (if they're jumping up). The way the energy is emitted or absorbed is in the form of photons of light, so the end result of lots of electrons jumping up and down between energy levels is a series of bright (emission) or dark (absorption) lines at characteristic frequencies within the spectrum.

These spectral lines are one of the most powerful tools available to astronomers. For one thing, they can be used to determine the chemical composition of an object that may be millions of light years away. But spectral lines have another use, too. When astronomers look at a star's spectrum, the relative placement of lines may tell them what chemical elements they're seeing, but if they measure the actual wavelength of those lines, they may find they're all slightly off by a fixed amount. This sounds like an annoying error, but it isn't – it's really useful information.

It comes down to the Doppler effect – something that many people will have heard of, but even if you haven't heard of it before, you will know about it without realising. Think about watching a motor race on TV; as the cars are approaching the camera position, they have a high-pitched sound, which abruptly changes to a low-pitched one as they whiz past and start to disappear into the distance. What's happening is that the sound waves are compressed as the cars approach, so their wavelength is reduced and their frequency (which we hear as pitch) is increased, and then when they recede, the waves are stretched out, resulting in a longer wavelength and lower frequency.

The same happens with light waves. If we look at a star that happens to be moving towards us, then all its spectral

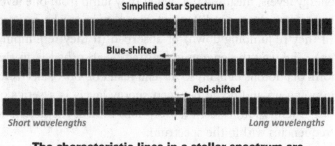

**Simplified Star Spectrum**

**Blue-shifted**

**Red-shifted**

*Short wavelengths*                    *Long wavelengths*

**The characteristic lines in a stellar spectrum are shifted to the blue when the star is approaching us, and to the red when it's moving away.**

NASA

lines will be shifted to higher frequencies, or towards the blue end of the spectrum. Conversely, if it's moving away from us, the lines are 'red-shifted' to lower frequencies. The amount by which the lines are shifted can be used to work out the exact speed of the star – or more correctly, the component of its velocity along the line of sight.

## The electromagnetic spectrum

When I refer to the 'red end' and 'blue end' of the spectrum, I assume everyone knows what I'm talking about. The sequence of colours produced by a prism, like the colours of the rainbow, run as follows: red, orange, yellow, green, blue, violet. In terms of wavelength, this covers a range from around 750 nanometres (deep red) down to 380 nanometres (extreme violet), or equivalently frequencies from 400 terahertz up to 790 terahertz.

But why just this range? What happens if we increase the wavelength to 800 nanometres, or the frequency to

800 terahertz? It's still possible to produce light-like waves – called infrared and ultraviolet respectively – but it's just that we can't see them. In the same way that objects of a certain temperature radiate light of different colours, objects that are even hotter or cooler produce radiation in parts of the spectrum that are invisible to us. The generic term for this is electromagnetic radiation, and it's what's emitted when electrons or other microscopic particles shift from one energy level to another. It's called 'electromagnetic' because the resulting wave takes the form of oscillating electric and magnetic fields – which belatedly explains what it is that's doing the 'waving' in a light wave.

The reason we only see a part of the electromagnetic spectrum is down to the way our eyes evolved. The visible wavelengths happen to be the ones at which the bulk of the Sun's energy reaches the Earth's surface – sunlight, in other words – so it makes sense that we've evolved to see the world around us as it's reflected at these wavelengths. But the Sun also produces plenty of infrared and ultraviolet radiation. We feel the former as heat, and can artificially 'see' at infrared wavelengths with the aid of night-vision goggles. As for UV, the fact that it's higher in frequency than visible light means that its photons carry more energy – enough to damage the cells of our bodies if we're exposed to too much of it. Fortunately, however, most of the ultraviolet radiation from the Sun is absorbed high up in the Earth's atmosphere, before it gets to us.

Even with the addition of infrared and ultraviolet on either side of the visible light we're familiar with, we haven't exhausted the whole of the electromagnetic spectrum. At longer wavelengths and lower frequencies than infrared, light takes the form of radio waves. Humans don't have any natural means of detecting radio frequencies, and we remained in blissful ignorance of their existence for millennia, but in the past 100 years it's become indispensable to the technology we use every day. Car radios, television, mobile phones, Wi-Fi, radar and microwave ovens all rely on radio waves of one frequency or another in order to work.

If we go in the other direction, to frequencies – and hence photon energies – higher than the ultraviolet band, then we come first to X-rays and then gamma rays. The energies here are potentially even more damaging than ultraviolet, so it's a good thing there's very little natural radiation at these frequencies at the Earth's surface. Again, they're things we're most familiar with through technology – ironically, given their harmful potential, often in the context of (carefully controlled) medical applications.

Why am I telling you all this? I'm sure you've already guessed the answer – because telescopes designed to observe these invisible parts of the electromagnetic spectrum can tell us just as much about the universe as traditional visible-band telescopes. Astrophysicist Rhodri Evans makes a useful analogy with the 88 notes of a piano keyboard. If the lowest note represents the lowest-frequency radio waves and the highest note the highest-frequency gamma rays, then the familiar visible part of the spectrum would correspond to less than a single key of the piano.

In hindsight, widening out the spectrum with which astronomers view the universe might seem like a no-brainer. Cooler stars than the Sun are going to shine brightest at infrared wavelengths, while hotter ones will be strongest in the ultraviolet. The cold gas of interstellar space will only be visible through its radio emissions, while high-energy

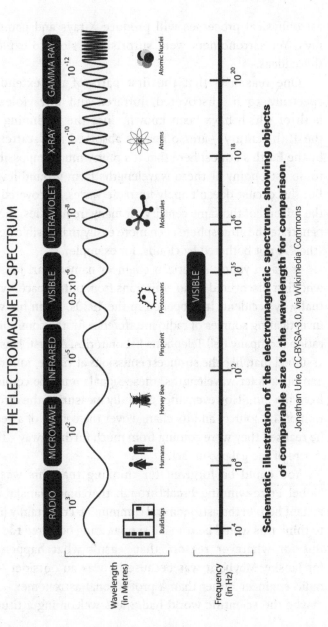

**Schematic illustration of the electromagnetic spectrum, showing objects of comparable size to the wavelength for comparison.**

Jonathan Urie, CC-BY-SA-3.0, via Wikimedia Commons

astrophysical processes will produce X-rays and gamma rays. Yet astronomers were surprisingly slow to exploit these ideas.

One reason is that the first parts of the extended spectrum to be discovered, infrared and ultraviolet – both of which have been known since the beginning of the 19th century – are so heavily absorbed and scattered by the Earth's atmosphere that it's pretty much impossible to do astronomy at these wavelengths from ground level. But this excuse doesn't apply for radio waves, discovered at the tail end of the same century. At many frequencies, these penetrate the atmosphere even more easily than visible light (they're not bothered by clouds, for example).

It was a young American engineer named Karl Jansky who first discovered radio emissions from outer space – virtually by accident. It happened in the 1930s, when he was investigating sources of radio interference for the communications company Bell Telephone Laboratories. At first, Jansky assumed that, like the strongest emissions at visible, infrared and ultraviolet wavelengths, these signals must be coming from the Sun. However, after carefully measuring the direction of the source, and its change over the course of a year, he realised they were coming from much further away – the centre of the galaxy, in fact.

You could be forgiven for thinking that this was a Nobel Prize-winning breakthrough that made Jansky an instant hero in the astronomical community. I'd certainly like to think this would be true if a similar thing occurred today. But for whatever reason, that wasn't what happened for Jansky. Maybe it was because he was an outsider – a radio engineer rather than a professional astronomer – or maybe the scientific world had a less welcoming attitude to 'new technology' in those days. The sad fact is that most astronomers were sneeringly dismissive of Jansky's discovery, and – aside from a few amateur dabblers – no one seems to have leapt at the idea of making a purpose-built 'radio telescope'. It wasn't until the 1950s that it finally dawned on astronomers that they really should take a closer look at radio wavelengths.

The results, within a few years of the first large radio telescopes being constructed, were staggering. Almost as many new astronomical phenomena were revealed through radio astronomy as had been discovered in three centuries of observations with optical telescopes. The invisible cold gas that permeates galaxies, and spreads out far beyond their previously known boundaries, suddenly became visible. Unimaginably powerful radio sources were discovered lurking at the centres of some galaxies, blasting out far more energy than can be accounted for by the visible stars there. Completely new objects, given fanciful names like pulsars and quasars, solved longstanding mysteries of stellar and galactic evolution – and raised many new mysteries of their own.

If astronomers had been slow on the uptake when it came to radio astronomy, it wasn't a mistake they were going to make twice. They knew they now had to look at the rest of the electromagnetic spectrum in just as much detail, as well. The timing of this realisation couldn't have been better, since we're talking about the end of the 1950s and the beginning of the 1960s, which just happened to coincide with the beginning of the space age. Aside from radio and visible wavelengths, astronomy in other parts of the electromagnetic spectrum – the infrared, ultraviolet, X-ray and gamma ray bands – is all but impossible to do from the Earth's surface, due to the cloaking effect of the atmosphere. So we've finally come face to face with the subject of this book – telescopes in space – and not as a some kind of luxury scientific development, but as a necessity when it comes to picking up electromagnetic waves from outer space.

Actually, a space telescope has important advantages over a ground-based one even at visible wavelengths. That's because, even though starlight can pass through the atmosphere, it doesn't do so unscathed. One of the most distinctive features of stars is that they 'twinkle', rather than appearing as steady points of light. This twinkling doesn't originate in the star itself, but in the Earth's atmosphere. What's happening is that the ultra-thin ray of light from the star gets jiggled around by turbulence as it passes through the atmosphere – which needless to say, is a bad thing for astronomers. It means that no matter how good a telescope's theoretical resolution, it's never going to form a perfectly sharp image even on a completely cloudless day.

The first person to propose the obvious solution – putting a telescope in orbit above the atmosphere – was an American physicist named Lyman Spitzer, who produced a short paper entitled 'Astronomical Advantages of an Extra-terrestrial Observatory' in 1946.* Describing a hypothetical future satellite consisting of a large reflecting telescope equipped with a variety of measuring devices for different types of astronomical research – which, in hindsight, sounds a lot like Hubble – he wrote:

---

* https://history.nasa.gov/SP-4407/vol5/chapter-3/III-1%20 (546).pdf

> If the figuring of the mirror could be sufficiently accurate,
> its resolving power would be enormous ... an object on
> Mars a mile in radius could be clearly recorded at closest
> opposition while on the Moon an object 50 feet across
> could be detected with visible radiation. This is at least ten
> times better than the typical performance of the best ter-
> restrial telescopes.

In his paper, Spitzer also pointed out the other key benefit
of a space telescope: that it would be able to see beyond the
visible spectrum into those parts of the infrared and ultraviolet that are blotted out by the Earth's atmosphere. At the
time, however, these insights were purely academic. It wasn't
until the end of the following decade that Earth-orbiting satellites became a reality, with the launch of the Soviet Union's
Sputnik 1 in 1957 and America's Explorer 1 the following
year.* The latter launch took place on 31 January 1958; less
than six months later, on 29 July 1958, the US government
created a special agency, the National Aeronautics and Space
Administration (NASA), to coordinate all its space-based
activities.

From the start, one of those activities was going to be
astronomical research, so the new agency needed to find
someone to run that side of the space programme. The
person they chose as their first chief of astronomy – and
probably the single most influential figure in the history of
space telescopes – was Nancy Grace Roman. In her thirties

---

* Despite its huge importance in the wider history of space exploration, the Soviet Union (and later Russia) had surprisingly little
involvement in space-based astronomy, which has always been dominated by the United States and, to a lesser extent, Western European
countries.

at the time, Roman had already made a name for herself both through work with ground-based optical telescopes and in the fledgling field of radio astronomy. Spending two decades in post, Roman guided NASA's astronomical programme throughout the 1960s and 1970s, well into the planning stages for what would eventually become the Hubble Space Telescope. For that reason, she's often referred to as the 'Mother of Hubble' – although that's really only the tip of the iceberg in terms of the projects she had a hand in.

Over her career, Roman worked on space telescopes across virtually the whole of the electromagnetic spectrum. At the high-frequency end, there was the Uhuru X-ray Satellite, which will feature prominently in Chapter 7. At lower frequencies, there was the self-explanatory Infrared Astronomical Satellite (IRAS) and a microwave telescope called the Cosmic Background Observer (COBE), both of which we'll meet in Chapter 3. Another focus of the early space telescope programme was the ultraviolet band, which is a part of the spectrum where our atmosphere is particularly impenetrable. So even a relatively tiny telescope up in space can show us things we have no way of seeing from down here on the surface.

In fact, the first space telescope of all, rather clunkily named Orbiting Astronomical Observatory 2, was optimised for ultraviolet. Overseen by Nancy Roman and launched in December 1968, this gave us our first real glimpse of the ultraviolet universe. It was followed a decade later by the International Ultraviolet Explorer (IUE) – a collaboration between NASA and the European Space Agency (ESA). In an interview, Roman later said this was the project she felt most proud of:

> And the reason I say that is that I think every other major
> project that I was involved with would have occurred
> sooner or later without me. I don't think IUE would have.*

Launched in 1978 and operated continuously for eighteen years – far longer than its planned lifetime of three years – IUE was space-based astronomy's first truly spectacular success. To start with, it was the first space telescope that came anywhere near the optical sophistication of a ground-based telescope, with a 45-centimetre mirror feeding data into a spectrograph and a TV-style camera. Unlike its predecessors, it could be operated by astronomers in real time, and its orbit was specially chosen to allow extremely long exposures to be taken. The data collected by IUE – amounting to some 100,000 observations – is still in use today and has contributed to over 4,000 peer-reviewed papers.

So there's no doubt that IUE made a tremendous impact on the scientific world. On the other hand, like pretty much every other telescope, whether on the ground or in space, the general public remained blissfully unaware of its existence. There's no surprise there; people, by and large, don't get very excited by pure science research – and that, after all, is what space telescopes are designed to do. The surprise came when one of them broke ranks and started talking in a language – visual imagery that often seems closer to art than science – that everyone could understand, regardless of their background. That telescope's name, of course, was Hubble – and it's the subject of our next chapter.

---

* https://www.aip.org/history-programs/niels-bohr-library/oral-histories/4846

# HUBBLE 2

'A game-changer for civilisation's perception of the universe and our place in the universe.'

That's how Ray Villard, the news director at NASA's Space Telescope Science Institute (STScI) in Baltimore, described the Hubble Space Telescope on the occasion of its 30th anniversary in 2020.[*] In a different context, this could be dismissed as typical PR hype, but in this case it's little more than a statement of fact. Astronomers have made amazing discoveries about phenomena from the atmospheric composition of extra-solar planets to the increasing speed at which the universe itself is expanding that weren't even dreamed of 30 years ago – and it's all thanks to data collected by Hubble.

It's not just professional astronomers whose view of the universe has been shaped by Hubble. If anyone, in any walk of life, has a mental image of the depths of space – or

---

[*] https://www.youtube.com/watch?v=V3xO0BEbCck

an actual image of a galaxy or nebula as the background image on their laptop or phone – then it's a pretty safe bet that image came from Hubble. Over the years, countless people, organisations and instruments have contributed to our understanding of the universe – and, with a few notable exceptions like Einstein and NASA, their names remain unknown to the public at large. So it's only people with a serious interest in astronomy who will have heard of Dr Edwin Hubble, the man who first observed the expansion of the universe, or the 100-inch (2.54-metre) Hooker telescope on Mount Wilson in California that he used to do it. Yet almost everyone has heard of the orbiting telescope that was named after Dr Hubble.

In the astronomical community itself, most astronomers can claim at least a tenuous connection to Hubble, even if it's just that one of their research papers drew on another paper that used data from Hubble, or they attended a conference where new Hubble results were presented. My own connection goes back to 1984, the year after the as-yet-unlaunched Hubble got its name (prior to that, it had just been the 'large space telescope'). During that summer and the following one, I spent a few months working at STScI, which had already been set up in advance of Hubble's launch. Today, it's the place where Hubble data is analysed; in those days its research was focused on areas that Hubble might be used to investigate.

My own work, in collaboration with an inspirational, larger-than-life Australian named Colin Norman, was published in 1985 as 'Black Holes and the Shapes of Galaxies'. While I moved on to various other jobs after that, Norman stayed at STScI – witnessing Hubble's launch from Cape Canaveral in 1990 and all its astonishing successes since then. For the 30th anniversary in 2020, I was asked to write a cover feature for *All About Space* magazine, to include a number of soundbites from scientists who had been involved with it over the years.* One of the people I approached was, of course, my old friend Colin Norman, who gave me the following quote:

> Hubble has changed the landscape of astronomy and astrophysics. It has far exceeded its early goals – no other science facility has ever made such a range of fundamental discoveries. It's been a privilege to be associated with this effort that has become embedded in the culture of our time.

I think that's a perfect summary of Hubble's success, spanning both its scientific achievements and its unique cultural impact. But those things did not come easily – they followed decades of painstaking planning and design that gained almost no publicity at all. So it's worth taking a moment to look more closely at how Hubble came into existence.

## NASA's great observatory

NASA's aspirations for a 'large space telescope' go back a long way. The US National Academy of Sciences endorsed the basic principle as long ago as the early 1960s, setting up a study group that was chaired by none other than Lyman

---

* A shortened version of the article can be read online: https://www.space.com/hubble-space-telescope-turns-30.html

Spitzer – the man who, two decades earlier, had first proposed the idea of an orbiting space observatory. As surprising as it sounds now, the natural assumption in those days was that any telescope in space would have to use either photographic film or a TV-style video camera to record images. It was only in the 1970s that the newly emerging technology of digital sensors started to look like a more practical alternative. The idea was championed by NASA's chief of astronomy, Nancy Roman – just one of many reasons why she later acquired the sobriquet 'Mother of Hubble'.

Roman also succeeded in toning down some of the more impractical proposals for the large space telescope, not least its size. She realised that it didn't actually need to be particularly large, compared to its terrestrial counterparts, in order to beat them hands down at their own game. Just as importantly, she realised that a telescope that looked affordable to US Congress – and therefore had a good chance of being built – was always going to be superior to one that was so expensive that it would never get off the ground. So the initial designs for a 120-inch (three-metre) diameter mirror were scrapped in favour of a smaller but more practical 94-inch (2.4-metre) one.

Another cost-saving measure was to make Hubble a collaborative project with the ESA, rather than a solo effort by NASA alone. It was agreed that ESA would supply Hubble's solar panels and one of its scientific instruments, in return for European astronomers receiving 15 per cent of the telescope's observing time. But NASA was still very much the senior partner, responsible for building the telescope itself – including the all-important main mirror – and launching it on board the Space Shuttle. In fact, Hubble was to be one of four 'great observatories' that NASA planned to launch with the

Shuttle, the others being the Chandra X-Ray Telescope, the Compton Gamma Ray Observatory and the Spitzer Infrared Telescope. We'll meet those other telescopes (the last of which ended up being smaller than originally planned, and not launched by the Shuttle) later in the book.

As mentioned in Chapter 1, Hubble uses a standard Cassegrain design in which two mirrors, a primary and secondary, are used to project the light collected from distant objects through a hole in the main mirror onto a sensor located behind this. In fact, Hubble allows the light to be channelled to one of several different scientific instruments. In the form in which it was originally built, there were five of these:

- Wide Field and Planetary Camera (WFPC): An optical camera covering the entire visible band and extending some way either side into both the infrared and ultraviolet, this was able to take photographs over a relatively wide field of view at the expense of having a somewhat lower resolution than Hubble's other sensors.
- Faint Object Camera: The instrument supplied by ESA, this had a narrower field of view than WFPC but higher resolution, effectively making it Hubble's 'telephoto lens'.
- High Speed Photometer: This wasn't a camera, but a relatively simple device intended to measure the changing brightness of variable stars and other fluctuating sources.
- Goddard High Resolution Spectrograph (GHRS): One of Hubble's two original spectrometers, this one was optimised for the ultraviolet band, and is capable of distinguishing very fine details in the spectra of the objects it analysed.

- Faint Object Spectrograph: Complementing the GHRS, this was able to examine fainter objects over a wider range of wavelengths, albeit with lower spectral resolution.

Hubble was originally scheduled for launch in 1986, but the tragic crash of the Space Shuttle Challenger at the start of that year threw NASA's schedule into disarray. After a number of further delays, Hubble finally made it into orbit on 24 April 1990, carried aloft in the payload bay of Space Shuttle Discovery. Unfortunately, the launch was almost immediately followed by disastrously bad news. As soon as the first images from the telescope were examined, it became clear that there was a serious flaw in the construction of the main mirror. Due to imperfectly calibrated equipment, it had been machined to a shape that was incorrect by about a thousandth of a millimetre. As small as that sounds, it meant that stars that should have been sharp pinpoints of light looked slightly blurred.

As we learned at the end of the previous chapter, one of the main reasons for putting a telescope in orbit is to avoid the blurring effect of the Earth's atmosphere. So, if Hubble's images were themselves going to be blurred, what was the point of it? Fortunately, the situation proved to be recoverable for two different reasons. First, the error in the shaping of the mirror wasn't due to an accident or shoddy workmanship. If that had been the case, it would have meant random inaccuracies that were very difficult to put right. But the people who made the mirror had worked with meticulous precision; the only problem was that they were working to the wrong specification. It's often said that Hubble's mirror was 'perfectly imperfect' – which turned out to be its saving grace. The imperfection was so precisely defined that it was a relatively simple matter to design an optical system to correct it.

The other reason NASA was able to recover from near disaster with the mirror stems from a basic feature of Hubble's design. Unlike most satellites, all along, Hubble had been designed for astronauts to be able to service and upgrade it in the course of subsequent missions. Thus, what had originally been intended as the first of several routine maintenance visits – by the Space Shuttle Endeavour in December 1993 – became a make-or-break rescue mission instead. If it had failed, the mission would almost certainly be remembered as one of the lowest points in NASA's history. But it succeeded – and now the mission isn't really remembered very well at all. But the plain fact is that – in my personal opinion, anyway – the Endeavour mission was the second greatest triumph of human spaceflight after the Moon landings. Without it, Hubble would have been doomed, and NASA would have wasted billions of dollars and decades of effort on a telescope that was no better than what already existed on Earth. As it was, Hubble was brought all the way back to its original design spec and survived to become the scientific and cultural icon we know today.

The repairs weren't easy. They involved five separate spacewalks, alternating between two pairs of astronauts, lasting a total of 36 hours. The main task as far as the mirror problem was concerned was to install a device called the Corrective Optics Space Telescope Axial Replacement (COSTAR). If you wear spectacles to compensate for your eyes' inability to focus perfectly, then that is basically what COSTAR did for Hubble. To make space for it, the

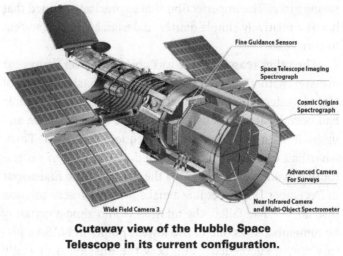

Fine Guidance Sensors

Space Telescope Imaging Spectrograph

Cosmic Origins Spectrograph

Advanced Camera For Surveys

Near Infrared Camera and Multi-Object Spectrometer

Wide Field Camera 3

**Cutaway view of the Hubble Space Telescope in its current configuration.**

NASA

cheapest of the telescope's instruments, the high-speed photometer, was removed. The astronauts also replaced Hubble's main camera, the WFPC, with an upgraded one, WFPC2. The latter had its own corrective optics built in, so unlike the other instruments, it didn't need to use COSTAR.

After Endeavour, astronauts returned to Hubble on more-or-less routine servicing visits on four further occasions, in February 1997, December 1999, March 2002 and May 2009. In the course of these – as well as fixing the occasional problem with ageing hardware – Hubble's various instruments were gradually upgraded to ensure it remained state-of-the-art. The new instruments, like WFPC2, had built-in corrections for the faulty mirror, so by the time of the final servicing visit in 2009, COSTAR was no longer needed. This meant it could be replaced with a brand-new instrument – an ultraviolet spectrometer called the Cosmic

Origins Spectrograph, designed to study faint, point-like sources – which finally brought Hubble's complement of scientific instruments back up to five.

## Hubble's view of the universe

There are two sides to Hubble, a public-facing one and a side devoted to cutting-edge scientific research, with limited overlap between the two. For example, the publicly released images for which Hubble is so famous are far from being pure science; they include a fair amount of artistic input that's been added by STScI's Office of Public Outreach.

Perhaps most surprising of all is the fact that Hubble doesn't take colour photographs. That's because it's first and foremost a scientific instrument, and scientists don't like working with colour images. What they do want are a whole series of black-and-white images, taken through filters that only let a narrow range of wavelengths through. This provides them with more hard data to work with than a colour image. To produce all those spectacularly colourful images for a wider audience, the Public Outreach team uses a mixture of scientific and artistic judgement to colour and stack the individual black-and-white photos to create a single composite image. While reflecting the true colours of the scene to some extent, these are also shaded in ways specifically designed to appeal to the human eye – often intentionally mimicking earthly landscapes we're already familiar with.

The first truly iconic image released by the Hubble team, in April 1995, was given the evocative name 'Pillars of Creation'. It was a masterpiece of scientific PR – one

**Hubble's second, even more spectacular, view of
the 'Pillars of Creation' dates from 2014.**

NASA

of the greatest in history – and couldn't have been more
timely as it came hot on the heels of the emergency repair
mission that finally brought Hubble up to its full capability.
The picture shows part of the Eagle Nebula, a star-forming
region 6,500 light years away in the constellation of Serpens.
This close-up shows a group of young stars surrounded by
brightly illuminated gas, as well as darker silhouettes formed
from interstellar dust. As impressive as the original 'Pillars of

Creation' image was, the Hubble team were able to improve on it – after additional shuttle visits upgraded the telescope's capability still further – when they revisited it in 2014.

Hubble's colourful images serve a valuable function by bringing the wonders of astronomy home to the general public, and in particular in helping to inspire young people to pursue an astronomical career. With a reputation that can be undeservedly stodgy, the physical sciences tradition- ally struggle to attract the best and brightest talent. In that context, pictures like the 'Pillars of Creation' are the best recruiting posters imaginable.

Even so, the fact remains that Hubble really only exists for one purpose, and that's to do science. In this respect, it's been as successful as any telescope in history. In over 30 years of operation, it's contributed to every branch of astronomy, from studies of our own Solar System to the most distant objects in the universe. Well over 15,000 scientific papers have been published describing the results obtained with it. Many of these don't rely on photographic images at all – since, as we've seen, Hubble has other instruments besides cameras. Its spectrographs can tell astronomers about the chemical composition of distant objects, and the speeds they're moving at.

This may not sound as exciting as a high-resolution photographic image of a galaxy or nebula, but in the right cir- cumstances it can be every bit as dramatic. This was certainly the case with one application of Hubble's spectrographs that could scarcely have been imagined at the time the tele- scope was launched. Back in 1990, no one knew for certain whether there were any planets orbiting stars other than the Sun. It seemed a reasonable theoretical possibility, but because they'd never been observed, it was a subject more often encountered in science fiction than professional astronomy. Within a couple of decades, however – as we'll see in more detail in Chapter 4 – the existence of such 'exoplanets' had been established beyond doubt, and astronomers started finding them in larger and larger numbers.

Finding exoplanets calls for specialised instruments, and it's not a task that Hubble is particularly well suited to. On the other hand, all that a specialist planet-hunting telescope can do is tell us that an exoplanet is there; it can't tell us other things we might want to know, such as what the planet's atmosphere is made of. This is where Hubble's spectrographs come in. Using a technique called transmission spectroscopy, these instruments can probe the atmospheric composition of exoplanets that have previously been located by another telescope. This involves analysing the light from the exoplanet's host star after it has passed through that planet's atmosphere.

So far, this all sounds rather academic, so you may be wondering why I said the results can be as exciting as any of Hubble's photographs. The key is when we find indications in an exoplanet's atmosphere of chemicals, such as water, which – here on Earth – we associate with the presence of life. As exciting astronomical discoveries go, it's hard to beat the one Hubble made in 2019, when it found clear indications of water on the Earth-like exoplanet K2-18b.

Careful analysis of Hubble's images can result in cutting-edge science too – even, ironically enough, in the study of invisible components of the universe, such as so-called dark matter. I refer to this as 'cutting-edge' because its mysterious nature is still a hot topic in astronomical circles, although the term itself dates all the way back to the 1930s. It was coined by Swiss–American astronomer Fritz Zwicky, who noticed an odd discrepancy between the visual appearance of the galaxies in a cluster and their relative motions within the cluster. In the same way that a spacecraft can attain 'escape velocity' from Earth's gravity if it's moving fast enough, there's a maximum speed a galaxy can travel at inside a cluster before it flies off at a tangent. Zwicky found that in order to provide sufficient gravity to hold all the galaxies in, the cluster had to contain significantly more mass than he could see visually. This invisible but gravitationally essential mass is what he meant by 'dark matter'.

But if dark matter is invisible by definition, how can you see it in a Hubble image? The answer is that while it can't be observed directly, it's possible to infer something about it using an effect called gravitational lensing. This is one of the consequences of Einstein's theory of General Relativity – a description of how space, matter and gravity work, which is so notoriously complicated that, even today, many physics graduates struggle to understand it. But shorn of its mathematical details, gravitational lensing is about as close to 'intuitively obvious' as Einstein's theory ever gets. It simply says that the path of a light ray is bent as it whizzes past a massive object due to the pull of its gravity.

This means that, just as light rays are bent by a telescope lens, they can also be bent by strong gravitational fields. If a distant, very bright object – usually a special kind of super-luminous galaxy called a quasar – appears close to the same line of sight as a dimmer, nearby galaxy, then the light rays from the distant object will be bent around the nearer one. Depending on the precise geometry, it's even possible for the distant object to be split into several separate images.

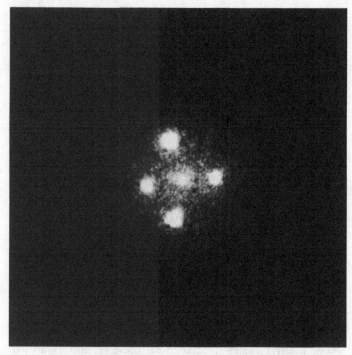

**Hubble's view of the 'Einstein Cross' – a distant
quasar that we see as four separate images due to
gravitational lensing by a closer intervening galaxy.**
NASA

In some cases, such as the appropriately named 'Einstein
Cross', the result can be a satisfyingly symmetrical pattern.

While gravitationally lensed quasars were known
before Hubble's time, the telescope has given us much
more detailed views of them, as well as discovering a host
of previously unknown ones. The relevance to dark matter
comes not from the quasar itself, but the intervening galaxy
because the precise way in which it distorts the quasar image
depends on the distribution of dark matter within it. So, by

comparing the Hubble image with the predictions of different computer models, astronomers can work out what that distribution must be.

The word 'dark' in dark matter is a good description of the current state of our understanding – or lack of it – concerning its nature, but not of its optical properties (if it was dark in the literal sense, it would blot out objects behind it and we could trace its distribution that way). Much the same is true of another cosmological mystery, and one that Hubble was instrumental in discovering: dark energy. This is closely related to the discovery for which the telescope's human namesake, Edwin Hubble, is most famous: the expansion of the universe. From some popular accounts, you may get the impression this is some kind of giant, ongoing explosion, with all the other galaxies flying away from our own. But that's a misleading picture. A much better analogy, which originated with Edwin Hubble himself, is a fruit cake baking in an oven. As the cake rises, every single raisin moves away from every other raisin, and the further apart they are, the faster they move away from each other. In greatly over-simplified terms, that's what's happening (with galaxies instead of raisins) when we talk about the expanding universe.

Given this picture, it seems reasonable to assume that the expansion of the universe is gradually slowing down, as it's pulled back by the combined gravity of all the matter in it. This gave astronomers a neat way to estimate the total mass of the universe, by measuring the rate at which its gravity slows down the expansion. It isn't the easiest measurement to make, but one way to do it involves a particular class of objects called Type Ia supernovae. The necessary observations were made by two separate teams, using

Hubble data gathered between 1995 and 1997 – but as good as the data was, it didn't help them calculate the mass of the universe. To do that, they needed to know the rate at which the expansion of the universe was slowing down, but the Hubble data didn't tell them that. It told them the expansion was speeding up.

The result was totally unexpected, and no one fully understands it to this day. It basically says that the universe behaves as if it's permeated by a mysterious something – dubbed 'dark energy' for want of a better term – that counterbalances the large-scale effects of gravity and pushes even harder in the opposite direction. It was an astonishing discovery, and one that resulted in the first ever Nobel Prize for a member of the Hubble team. STScI scientist Adam Riess was awarded a share of the 2011 physics prize for his part in 'the discovery of the accelerating expansion of the universe through observations of distant supernovae'.

When I was talking about the various advantages of orbiting telescopes over ground-based ones, I missed one out. The atmosphere not only blurs any view of space, but it also means the sky seen from Earth never gets totally dark. This is obvious enough in towns and cities that suffer from light pollution, but there's some residual skyglow even on an isolated mountaintop. You can visualise this most easily if you think of the daytime sky, which is illuminated from horizon to horizon due to the scattering of sunlight by the atmosphere. The same kind of scattering, even if it's just of starlight, happens at night too. At Hubble's altitude, on the other hand, the sky in any direction away from the Sun is completely black, so it can make out unbelievably faint objects if it uses a long enough exposure.

Hubble's longest exposure photographs, taken by repeatedly pointing at the same patch of sky over a period cumulatively amounting to many days, are called 'deep field' images. The name comes from the fact that the longer the exposure, the deeper into the universe the telescope is seeing. It's also seeing further back into the past, due to the finite speed of light. This means the further away an object is, the longer ago its light started out on its journey to us.

This makes Hubble something of a cosmological time machine. In this context, its most impressive result is

**A version of the Hubble Ultra Deep Field released in 2012, showing the most distant galaxies observed to date.**

NASA

the Ultra Deep Field image created from a cumulative 23 days of observing time spread out over the decade between 2002 and 2012. The image relates to a small patch of sky in the constellation of Fornax, specially chosen because it's virtually free of nearby objects that might confuse the picture. The result – containing over 10,000 individual objects, most of them galaxies – is like an enormously long core sample taken through that part of the universe, with increasingly faint and distant objects being further and further back in time.

Hubble's deep-field images are among its greatest achievements, pushing our view of the universe back towards its very beginnings. Another of the astronomers I spoke to for the 30th-anniversary *All About Space* feature was Garth Illingworth, a professor at the University of California who has been closely involved in Hubble's studies of the early universe. This is what he had to say about Hubble's impact:

> Before Hubble, we knew essentially nothing about galaxies in the first half of the life of the universe, i.e. in the first 7 billion years of the universe's 13.7 billion year life. Hubble has absolutely revolutionised our understanding of how galaxies grow and develop in the universe in that 7 billion years. *

In the context of one particular galaxy, GN-z11 in the constellation of Ursa Major, Illingworth pointed out that Hubble is looking back through 97 per cent of all time to see it as it was just 400 million years after the Big Bang. That's something

---

* The words I'm quoting here are from Professor Illingworth's original email to me; they were edited down for magazine publication.

that couldn't possibly be done with ground-based telescopes, which are incapable of seeing such a faint object against the skyglow of Earth's atmosphere.

In any other context, 400 million years might sound like a long time, but you have to think of it as a fraction of the age of the universe. If you liken the universe now to, say, a 50-year-old middle-aged adult, then Hubble is seeing GN-z11 as it was when the universe was an eighteen-month-old toddler. But what if we push that back more than a thousand times further, to a point just 380,000 years after the Big Bang when, in our human analogy, the universe was a half-day-old newborn baby? That's far beyond Hubble's capability, but there are other telescopes, operating in a different part of the electromagnetic spectrum, that really can see that far back in time. We'll meet them in the next chapter.

# PROBING THE BIG BANG 3

We live in an evolving universe: one that began at a finite moment in time, just under 14 billion years ago. At that point, all the matter we now see spread out in billions of galaxies was condensed into an unimaginably tiny volume, and it's been expanding in size ever since. This view has become so predominant among scientists, and has permeated the news and popular media so thoroughly, that I suspect most people today take it for granted. But that hasn't always been the case, even among scientists.

There's a hint of this in the name the theory is known by – the 'Big Bang' – which, if you think about it, is a rather flippant way of describing such an awesome event. In fact, the term was coined by the British astrophysicist Fred Hoyle during a 1949 radio debate, to describe a theory he disagreed with. There's a distinct pre-echo of it, too, in words that Hoyle's equally sceptical compatriot Sir Arthur Eddington wrote in 1927:

> The notion of a beginning is repugnant to me ... I simply do not believe that the present order of things started off with a bang.

Eddington was referring to a recent development in the then-new field of General Relativity – Einstein's notoriously complicated theory of space, time and gravity that we met in the previous chapter in the context of gravitational lensing. In the early days, its biggest attraction to scientists was that it gave them a mathematical framework for describing the large-scale behaviour of the universe as a whole. One of the simplest solutions of Einstein's equations – the one that Eddington objected to – looks very much like the Big Bang picture. But it's not the only possible solution, and Eddington felt the real universe had to be eternal and, at least when viewed on large scales, unchanging.

You might think that Edwin Hubble's discovery that the universe is expanding, which he announced in 1929, clinched the matter in favour of the Big Bang. But it wasn't that simple. As paradoxical as it sounds, it's possible to have a universe that's eternal and unchanging, as Eddington wanted, and in a continuous state of expansion, as Hubble observed, both at the same time. This is the situation in the so-called Steady State model of the universe, which is an alternative solution of Einstein's equations that was championed by Fred Hoyle and others. The only trick is that it requires the ongoing creation of new matter – equivalent to just a few hundred atoms per year per galaxy – so as to maintain a uniform average density as the universe expands.

Throughout the middle decades of the 20th century, both models – the Big Bang and Steady State – were equally consistent with observations. Today, that's no longer true, and we know for certain that the universe has evolved over the past several billion years in just the way the Big Bang model predicts. We touched on one piece of evidence at the end of the previous chapter, when we saw how the Hubble telescope has shown us galaxies that are far more distant – and hence seen at much earlier times – than anything that had been observed before. These earliest galaxies are noticeably different from the more mature ones we see at closer distances, in the sense that they tend to be smaller and more irregular-looking.

That's a relatively modern discovery, however. By the time Hubble's deep-field images were obtained, there had been little doubt about the Big Bang theory for several decades. The real clincher was the discovery of something that predated even the first galaxies – or the first atoms, for that matter. This is called the cosmic microwave background (CMB), and it's something that couldn't possibly have existed in the sedately unchanging universe of the Steady State theory.

Journalists often describe the CMB as the 'echo of the Big Bang', which makes it sound much more straightforward and easy to grasp than it really is. To start with, the Big Bang didn't make a sound because – in those very first moments – there wasn't any material substance for sound waves to travel through. When Fred Hoyle put the 'bang' into the Big Bang, he was simply using it as a metaphor for the abrupt creation of the universe out of nothing. But even if we can't hear the Big Bang, might we be able to see the flash of light from it, using a telescope that is even more powerful than Hubble?

Sadly, even this isn't possible. The newly formed universe wasn't just minuscule in size, it was also unimaginably hot. Even when the first material particles formed, you have to picture them something like the glowing, super-hot plasma that makes up the Sun. And no telescope can see inside the Sun because the particles in the plasma scatter light waves in a similar way to the scattering of sunlight by clouds in the Earth's atmosphere. It's the same when we try to peer further and further back in time towards the Big Bang. There comes a point – corresponding to the base of the clouds in the atmospheric analogy – beyond which we simply can't see.

The technical term for this fuzzy veil at the edge of the universe is the 'surface of last scattering', and it was formed around 380,000 years after the Big Bang. The radiation from it was emitted at a temperature comparable to that of a red dwarf star, around 3,000° Celsius – but the light from it doesn't look red to us today. It's become so stretched out by the expansion of the universe that its peak wavelength has increased from around 900 nanometres to roughly one millimetre, putting it in the microwave region of the electromagnetic spectrum.* It's this radiation (rather than any journalistically cosy 'echo of the Big Bang') that forms the observed CMB.

The existence of the CMB is a fairly obvious consequence of the Big Bang – obvious, that is, to anyone who delves deeply enough into the mathematics of it. So it was predicted theoretically – by Robert Dicke of Princeton University in New Jersey – before it was actually discovered observationally. That discovery, when it occurred soon afterwards, wasn't

---

* If you were paying close attention in the first chapter, you'll notice that the original wavelength of 900 nm puts it in the infrared part of the spectrum. This is indeed where the peak output of 'red' stars generally occurs, but we see them as red for the simple reason that we can't see infrared. The total emitted spectrum actually covers a wide range of wavelengths, as we'll see later in this chapter.

directly related to Dicke's prediction at all but came about more or less by accident.

It was an accident with curious echoes of an earlier one that we encountered in Chapter 1, when Karl Jansky unexpectedly discovered radio emissions from the centre of our galaxy while he was investigating sources of radio interference for Bell Telephone Laboratories in the 1930s. Around 30 years later, Arno Penzias and Robert Wilson were doing much the same thing – again for Bell, but with greater sensitivity, and at shorter, centimetre-scale wavelengths. To their mystification, they found a constant low level of background radiation, coming from all directions at once, that they simply couldn't explain or eradicate – even by clearing out some pigeons that had nested inside their huge, horn-shaped antenna and fouled its surface.

In one of the ironies of science history, Penzias and Wilson's antenna – now a carefully preserved national landmark – was located at Holmdel in New Jersey, just 25 miles from Dicke's team in Princeton – yet both groups were blissfully ignorant of each others' work. It was only when a mutual acquaintance, Bernard Burke, introduced them to each other in 1964 that the connection was made – and it was realised that the CMB had finally been found.

It was a discovery that came as close as anything to confirming the reality of the Big Bang, and definitively killing off its one serious rival the Steady State theory. It won the Nobel Prize for its discoverers, Penzias and Wilson – although not, to the annoyance of theoreticians everywhere, for the person who predicted it, Robert Dicke. Even so, it was just the beginning. A ground-based antenna might have been able to discover the CMB – but to study it in any depth was going to require a space telescope.

## The Cosmic Background Explorer

I've spent quite a lot of time introducing the CMB because although it's one of the most important topics in this book it may also be one of the least familiar. Taken out of context, it may seem a relatively dull subject for a space telescope to investigate. How can something that looks virtually the same in all directions – and that's seen at microwave rather than optical wavelengths – compete with colourful photographs of planets, nebulae and galaxies? The answer is that it's an utterly unique resource, the only practical method that humans have found so far by which they can directly study cosmology.

That's a word I used earlier in the book without properly defining it. Basically, cosmology refers to our understanding of the universe as a whole – how it originated, how it evolves and how it's structured – as opposed to the individual components of it, such as stars and galaxies, which are the subject matter of astronomy. As areas of study, both astronomy and cosmology have been found in all cultures, going back to ancient times. Traditionally, however, astronomy has always been a practical subject, based on direct observations of the night sky, while cosmology has been entirely theoretical. In the absence of any meaningful data, the earliest theories of the universe's origin were based purely on human imagination. Later, even those that had a firm grounding in science, such as the original Big Bang theory as developed from Einstein's equations of General Relativity, were still fundamentally speculative.

Edwin Hubble's discovery of the expansion of the universe, and his namesake telescope's observations of very distant galaxies, provided supporting evidence for the Big

Bang – but it was only indirect, circumstantial evidence. If you want to do any kind of direct observational cosmology, then there's only one way to go about it, and that's by taking a very close look at the CMB. After millennia of cosmological speculations, it's the nearest we're going to get to actually glimpsing the birth of the universe. And, as I said a moment ago, it's something that can only be done from space.

The basic problem is that the CMB signals we're looking for are very weak. This means that Earthbound telescopes are always going to struggle to detect them against mounting competition from human-generated microwave sources. The most obvious of these are microwave ovens, which operate at a wavelength of around twelve centimetres, but other technologies encroach even further into the wavebands relevant to CMB studies. The original discovery by Penzias and Wilson was made at around seven centimetres, in exactly the same part of the electromagnetic spectrum as modern 4G and 5G phones. High-bandwidth Wi-Fi makes use of even shorter wavelengths, of just a few millimetres, which is exactly where theoreticians expect the peak of the CMB radiation to occur.

The fact that the signal we're looking for is very weak has another consequence besides the potential for interference. Detecting it requires what radio astronomers call a 'long integration time', which is the equivalent to a long-exposure photograph in the optical world. In other words, the telescope has to stare at the same spot over a long period of time in order to collect a useful amount of data. This is another strong argument for putting a telescope in space, where it's able to point in the same direction continuously, without being hindered by the Earth's rotation. The frigid temperatures of outer space help too, because they provide better stability for the telescope's electronic components.

The first proposals for a space telescope designed specifically to study the CMB were submitted to NASA in the mid-1970s, around the same time that serious planning for Hubble got underway. The design that emerged became known as the Cosmic Background Explorer, or COBE for short. If the last word in its name seems to carry an echo of America's very first satellite, Explorer 1, that's no coincidence. Although it's rarely made headlines since 1962, NASA's Explorer programme – focused on small-to-medium-sized satellites designed for scientific research – has quietly continued to this day. Technically, COBE was Explorer 66 – and the International Ultraviolet Explorer, mentioned in Chapter 1, was Explorer 57. We'll meet several other members of the Explorer programme later on in the book.

Like Hubble, the original intention was that COBE would be launched using the Space Shuttle – for the simple reason that, in those early days, NASA was planning to do virtually all its space launches using the Shuttle. So COBE was designed with the Shuttle's relatively generous payload capacity in mind. Its launch was pencilled in for 1988, and development started in earnest in 1983. By that time, it had been decided that COBE would have a total mass of five tonnes and a diameter, when folded up for launch, of around 4.6 metres. As 'small-to-medium' satellites go, that's pretty hefty.

With the disastrous crash of Challenger in January 1986, COBE's goalposts moved – and in a big way. Unlike high-profile missions like Hubble, its Shuttle slot wasn't simply shifted back a few years – it was cancelled altogether.

After the COBE team had recovered from the shock, they realised it wasn't the end of the world because there are other heavy-launch operators besides NASA. They worked out that two available rockets, the Soviet Proton and European Ariane, were both capable of launching COBE into its desired orbit without changing its specification by a single gram or millimetre.

Today, perhaps, NASA bosses wouldn't blink an eyelid at such a suggestion because the competitive nature of the space launch sector – with numerous private companies vying for business – has become an established fact of life. But it was different in those days, and the NASA management team collectively blew its top. The COBE team were told in no uncertain terms that they could either launch on a NASA vehicle – anything but the Shuttle – or not launch at all.

In the time-honoured tradition of bureaucrats everywhere, this created a huge (and totally unnecessary) headache for the engineers because the only other launch rockets NASA had at the time were puny. The best that was available, the Delta 5000 series, would require shrinking COBE's design down to just 2.2 tonnes in mass and 2.4 metres in diameter. To make matters worse, NASA's deputy associate administrator* wanted a spacecraft delivered to the new specification within 24 months. Given that most space programmes tend to proceed at a glacial pace, this presented a serious challenge for the engineering team. Inevitably, the timescale slipped a bit, but – after one of the fastest development turnarounds in spaceflight

---

* This isn't a joke; there really was a person with this title, and his name was Sam Keller.

history – the slimmed-down, redesigned COBE was finally launched in November 1989, not much later than the originally planned Shuttle launch.

COBE was placed in a specially chosen orbit 900 kilometres above the Earth that was aligned almost perpendicular to the equator. This had several advantages, including putting it well clear of interfering signals from terrestrial sources and allowing the spacecraft's instruments to be oriented in a way that kept them permanently shaded from the Sun. Over the course of a year, it also gave the satellite a full 360-degree view of the entire sky, which would allow it to create a complete map of the CMB. This, indeed, was one of the main purposes of the mission, which we'll come back to in a moment. First, however, COBE had an even more crucial task to carry out.

There's an important aspect of the science of the CMB that I've rather glossed over so far. I said that the Big Bang theory predicts that, at a certain point in the distant past, the universe was filled with radiation similar to that produced by a red dwarf star, and that by the present time, this radiation will have been stretched out – in terms of wavelength – to the extent that it appears in the microwave band instead. I also said that Penzias and Wilson discovered microwave radiation at one particular wavelength, around seven centimetres, that's consistent with this theory. But if you think about it, this one measurement, together with a few other spot measurements that were made subsequently by ground-based radio telescopes, don't completely *prove* that what's being observed is the CMB. As far as credible theories go, it's by far the most likely explanation, but there's still a remote possibility that it might be some other cosmic source of microwaves that happens to emanate equally from all parts of the sky.

The solution to the dilemma is to make measurements, not just at two or three different wavelengths, but over a whole range of them to deduce how the relative strength of the observed radiation varies when it's plotted against wavelength. If the observations do indeed relate to the CMB, then according to the Big Bang theory, the resulting graph should take a very specific form known, rather obscurely, as a 'black-body spectrum'. This term derives from the fact that if you have a chunk of material that's completely black – in the sense that it absorbs every single photon, of whatever wavelength, that falls on it – and heat it to a given temperature, then the energy it radiates will follow a particular, well-defined mathematical law. In the case of the CMB, its black-body spectrum should correspond to the temperature of the universe at the time of last scattering – which was around 3,000° above absolute zero, or 3,000 Kelvin – reduced by a factor of around 1,000 to account for the stretching of the wavelength by the expanding universe. In other words, we're looking for a black-body spectrum somewhere around three Kelvin.*

This gives us a firm theoretical prediction of what the CMB spectrum ought to look like. As a bonus, it just so happens that there's no physical object in the known universe that, when heated to the specified temperature, produces a perfect black-body spectrum. Many objects come close, but they all deviate from that precise mathematical form. So if

---

* A quick reminder for anyone who may be unfamiliar with temperatures measured in Kelvin. The laws of physics put an absolute lower limit on temperature, which is around −273.15° Celsius. Scientists often find it easier to use this temperature as the zero of the scale – called absolute zero – and 'Kelvin' refers to values on this scale.

COBE found a perfect black-body form when it measured the CMB spectrum, it would be the final clinching proof that it really did originate at the surface of last scattering, 380,000 years after the Big Bang.

The first public announcement of COBE results was made at a meeting of the American Astronomical Society in Washington DC in January 1990, a mere two months after the satellite had been launched. In front of an audience of over 1,000 people, senior NASA scientist John Mather showed a graph of the hoped-for black-body spectrum, scaled to a temperature of 2.725 Kelvin, and then superimposed the COBE results on top of it. They matched perfectly. It was one of the great moments in the history of science – although one that, as far as I can tell, wasn't captured on camera. The

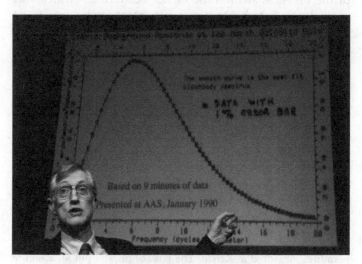

**NASA scientist John Mather with the results that showed a perfect match between COBE results and theoretical predictions for a black-body spectrum.**

NASA

picture included here is a kind of 'historic reconstruction' that NASA issued in 2006.

COBE's second major task, which was going to take considerably longer than the first, was to create a high-precision map of the CMB across the whole sky. The purpose of this was to test another prediction from the theoreticians, and I'm afraid it involves another piece of obscure scientific jargon. Hot on the heels of the black-body spectrum, we're now going to come face-to-face with 'anisotropy' – which, unless you're a scientist or a crossword addict, probably isn't a word you've ever had much occasion to use.

For better or for worse, I'm in the minority here because not only do I know what anisotropy means, but it's a word that elicits a decidedly negative emotional response in me. It goes back to a time that predates COBE, when I was doing the research for my PhD. This was related to the structure of galaxies, and it involved running a large simulation program on a Cray supercomputer. As sophisticated as it was for its time, the program had a number of weaknesses, not least in the way it represented the force of gravity. In the real world, this pulls equally in all directions – or, in science-speak, it's 'isotropic'. In my computer program, because of the approximations it used, gravity was frustratingly anisotropic. So I had to spend a significant part of that three-year project, and a whole chapter of my PhD thesis, trying to eradicate that expletive deleted anisotropy.

Anyway, the point is that anisotropy means 'not the same in all directions'. This may seem a strange idea to raise in the present context, because several times in this chapter I've made the point that the CMB, both theoretically and observationally, is pretty much the same in all directions – isotropic, in other words. Up to a point, this makes sense. The CMB is like a glimpse into the proto-universe, from which everything we see today evolved. And what we see today is, on the largest scales, pretty isotropic; whichever direction you look in, there's roughly the same number of galaxies of roughly the same shapes and sizes. At a finer level of detail, however, the universe certainly isn't the same everywhere. The mere fact that galaxies exist, and that they clump together into clusters and superclusters, shows us that much.

Cosmologists had always been pretty sure there must be some kind of primordial anisotropies in the CMB; the only question was on what scale, and whether they were large enough for COBE to detect. This was never going to be an easy task. Not only were the sought-for fluctuations likely to be very small, but there were other, much larger effects that had to be subtracted out of the data first. First, there were microwave emissions from astrophysical sources within our own galaxy, and on top of that a kind of illusory anisotropy caused by the motion of our galaxy relative to the CMB itself. But even in the face of these challenges, the COBE team were able to announce their findings after little more than a year of operation, in April 1992. They had indeed found primordial anisotropies in the CMB, at a minuscule level amounting to less than 0.001 per cent of its background value.

This was a huge breakthrough for the nascent science of observational cosmology, and a crucial first step in understanding the earliest history of large-scale structure in the universe. But some people went further over the top than others. Stephen Hawking, for example, speaking on TV in the immediate wake of the announcement, called it 'the scientific discovery of the century, if not of all time'. There are doubtless many scientists working in other fields who would dispute that claim, but if you take a really big-picture view, then maybe Hawking wasn't that wide of the mark. 'All time' encompasses not just modern science but the philosophers and prophets of the ancient world too. Surely, they would have given anything for a glimpse of the cosmos in the process of creation? Without too much exaggeration, that's what COBE's detailed measurements of the CMB provided us with.

If we think of COBE as giving us a glimpse of creation, how about a full-blown, high-resolution map of it? That's what COBE's successors, one from NASA and one from ESA, set out to provide.

## The view from L2

COBE was a hard act to follow. Its contributions to basic science far outstripped any space mission before it – a fact that was recognised in 2006 when it became the first spacecraft in history to win the Nobel Prize in physics. Or near enough, anyway; the prize actually went to two members of the COBE team – NASA's John Mather and his colleague George Smoot from the University of California – for, as the official citation reads, 'their discovery of the blackbody form and anisotropy of the cosmic microwave background radiation'.

NASA started work on a successor to COBE in 1996, by which time the latter had completed its mission. Another member of the Explorer series, the new spacecraft was formally designated Explorer 80, but it soon became known by the rather clever name of MAP. As an acronym for Microwave Anisotropy Probe, that's a good description of its purpose for anyone who understands what anisotropy means, while the word 'map' alone is close enough for those who don't. Sadly, however, one of the key members of the MAP science team, Dave Wilkinson, died soon after it was launched in 2001, and the mission was renamed Wilkinson Microwave Anisotropy Probe, or WMAP, in his honour. To avoid confusion, I'll use this form even when I'm talking about events before the name change.

WMAP had a much more compact design than COBE, with a mass not much more than 800 kilograms – far less than COBE's 2,200 kilograms, let alone the 5,000 kilograms of the original, Shuttle-launched design. This was partly because it was intended to be a simpler, lower-cost spacecraft, but in large part also due to advances in electronic technology in the decade between the two missions. In fact, WMAP was built to a much higher specification than COBE, with 45 times greater sensitivity and 33 times better spatial resolution. On top of that, it was designed to operate much further away from the Earth to minimise contamination from terrestrial emissions and provide a more stable thermal environment.

To understand just where WMAP was located, and why that particular spot was chosen, we have to turn to a peculiarity of orbit theory called the three-body problem. With just two bodies to worry about, for example, a satellite orbiting close to the Earth, then the mathematical solution has been known since the time of Johannes Kepler and Isaac Newton in the 17th century. In this case, the satellite's path around the Earth is a kind of distorted circle known as an ellipse. But if the satellite is sufficiently far away from the Earth that the pull of the Sun's gravity also has to be taken into account, the situation gets a lot more complicated. That's what the three-body problem is all about.

To cut a long story short, there are five points in and around the Earth's orbit about the Sun where the gravitational pull of the two bodies effectively cancel each other out. This allows a spacecraft to sit at one of these points – or, more practically, oscillate to and fro around it – virtually undisturbed. They're called Lagrange points, after Joseph-Louis Lagrange who studied them in the 18th century, and they're generally known by the abbreviations L1 to L5.* The one that's of particular interest to the designers of space telescopes is L2, which lies around 1.5 million kilometres further out from the Earth on a continuation of the line between the Sun and our own planet.

There are several reasons why L2 is a perfect location for a space telescope. It's far enough from the Earth that it isn't swamped by terrestrial emissions but close enough that communication is relatively straightforward. Perhaps most important of all, when you want to keep your electronic components as cold as possible for maximum stability, the two strongest sources of heat – the Sun and the Earth – always lie in exactly the same direction, so you only need a one-sided heatshield to keep the spacecraft cool.

WMAP was the first spacecraft to be sent to L2 – but far from the last. We'll meet several of the others later in the book, a couple of them before we even reach the end of this chapter. Strictly speaking, these L2-dwelling spacecraft aren't 'satellites' because they don't orbit the Earth. So I won't use that term for them – although it's a pretty minor

---

* These five points exist in any system that is dominated by two gravitating masses. We're interested in the Sun–Earth system here, but you may have come across them in the context of the Earth–Moon system as well.

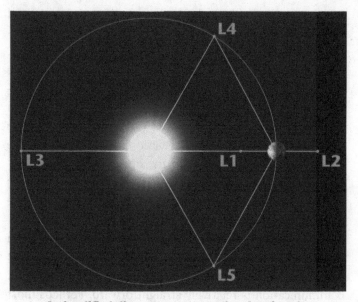

**A simplified diagram, not to scale, showing the
five Lagrange points of the Sun–Earth system.**
NASA

distinction that not everyone makes, since there really isn't
much difference either in functional or engineering terms.

Following its launch in June 2001, WMAP arrived at
L2 in October that year. It then embarked on a mapping
programme that continued for nine years, measuring
the intensity of the CMB over the entire sky once every
six months. The end result was a much higher precision
map than the one COBE had obtained – but it wouldn't be
long before this too was overshadowed by an even better one.

This was the work of the Planck telescope, which unlike
COBE and WMAP was operated by the ESA rather than
NASA. It was named after the German physicist Max Planck,
who is probably best known as the 'father of quantum

theory'. But that wasn't the reason he had this particular spacecraft named after him; it was because he was the originator of the 'black-body spectrum' formula that played such an important role in the discovery of the CMB.

There's a tendency to think that the biggest achievements in space are always NASA's, while ESA and the various other space agencies around the world play a humbler role. While there's a degree of truth in this – particularly in some of the more headline-grabbing areas of space exploration – it's not always the case, and several of ESA's science-focused missions have outstripped anything NASA has done in the same field. Planck is a case in point.

ESA began working on ideas for Planck around the same time as NASA was developing WMAP, in the mid-1990s – but all along it was a bigger and bolder concept. While its scientific aims were similar, its engineering specifications were higher, with more sensitive detection technology, better spatial resolution by a factor of 2.5, and a wider coverage of different wavelengths. The result was a much more expensive mission than its NASA counterpart, costing around $1 billion in comparison to WMAP's $150 million. The Planck mission also involved a much larger team than WMAP – although cynically, one might suspect this had as much to do with international politics as science. ESA missions are paid for by all its member nations, so it's not surprising that they all want a piece of the scientific action in return.

Planck was blasted into space on 14 May 2009 from ESA's spaceport in French Guiana, on board an Ariane 5-ECA launch vehicle. With a mass of around two tonnes, Planck was only a fraction of the full payload capacity of that rocket – and there was a good reason for that. It wasn't even the primary payload that day – an honour that went to the

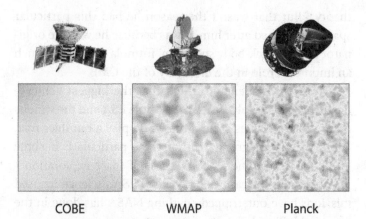

COBE          WMAP          Planck

**Small segments of the CMB maps created by COBE, WMAP and Planck, showing how the resolution improved each time.**
NASA/JPL-Caltech/ESA

3.4-tonne Herschel Space Observatory, which at the time was the largest infrared space telescope ever launched.* Herschel, like Planck, was on its way to L2, where both remained from mid-2009 until 2013 (as mentioned earlier, spacecraft don't literally sit at L2 but oscillate around it, so there can be several there at the same time).

In its more than three years of operation, Planck completed five all-sky maps, resulting in the highest-resolution view of the CMB to date. The above snapshots show how the quality of these maps has progressively improved through the three generations of spacecraft, from COBE to WMAP to Planck. While even the newest of these views may not have the visual impact of a Hubble photograph, the data it

---

* It's since been beaten by NASA's James Webb Space Telescope, which we'll meet in Chapter 6.

represents is invaluable to cosmologists, allowing them to fine-tune their models of the early universe in order to give us a more detailed understanding of its origins.

If there's any other astronomical topic that can compete with the origin of the universe in terms of its sheer fascination for humans everywhere, then it's the idea that there might be life on other planets besides Earth. Before the days of space telescopes – in other words, until very recently – the idea of potentially habitable, Earth-like planets beyond the Solar System was almost exclusively a subject for philosophers and science fiction writers. But that's all changed now, as we'll see in the next chapter.

# EXOPLANET HUNTERS 4

Among all the branches of astronomy, the search for exoplanets is one of the newest. The subject of planets orbiting stars beyond our own Solar System has really only become a subject of serious study in recent years. We touched on it briefly in Chapter 2, when we saw how the Hubble telescope has been used to probe the composition of exoplanet atmospheres. As far as Hubble was concerned, however, this only came as a belated afterthought since the first exoplanet hadn't even been discovered when it was launched in 1990.

Several different methods are now used regularly to locate new exoplanets, but I'm going to focus on just one of them that's particularly well suited to space-based telescopes. It's called the transit method, and it hinges on a principle that – as astrophysical theories go – is really very easy to grasp. When a planet orbiting a distant star passes in front of that star, as seen from an observer's perspective, then it blocks out a tiny proportion of the star's light. So, to detect the planet, we just have to watch for that characteristic dip in the star's brightness.

Monitoring the brightness of a star as a function of time is nothing new in astronomy. Many stars display variable brightness, and the study of their 'light curves', as they're called – essentially just a graph of brightness versus time – is a well-established technique. The only catch, in the context of exoplanets, is that the dip we're looking for is a very small one. If you turn the situation around, and imagine astronomers on an exoplanet observing our own Solar System, then its largest planet, Jupiter, would produce a dip of just 1 per cent when it transited the Sun. In the case of the Earth itself – which, of course, is more representative of the kind of planets we're most interested in finding – the dip would be a mere 0.01 per cent.

The transit method doesn't always work. If the plane of an exoplanet's orbit is tilted at a large angle to our line of sight, then we'll never see it pass in front of its host star. Even in cases where the planet does make transits, they may be few and far between. A strategically placed alien astronomer might hope to see the Earth make a transit once

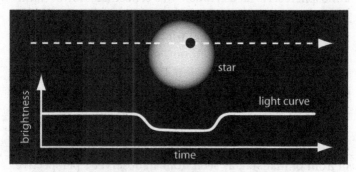

**The transit method of exoplanet detection depends
on measuring the small dip in a star's light curve
when one of its planets passes in front of it.**
NASA

per year (since that's the time it takes us to complete an orbit of the Sun), but this interval rises to twelve years in the case of Jupiter – and it may be necessary to observe several transits before a detection can be fully confirmed. If you add these problems to the smallness of the effect we're looking for, you could be forgiven for thinking the transit method is doomed to failure. Yet it's proved astonishingly successful – thanks in large part to a couple of specially designed space telescopes.

## Kepler

If you pick a star in the sky at random and monitor its light curve for a few hours, it's highly unlikely you'd be lucky enough to observe a transit. If you monitored it continuously for a couple of years, then – if it does indeed have a planet and the geometry of its orbit is just right – you may see a transit. But you'd just have one detection to show for many months of observation. What you really need to do is monitor a huge number of stars – continuously and simultaneously – and analyse all their light curves in search of that tell-tale little dip in brightness.

Even today, when framed in this way, the endeavour sounds utterly crazy, so it's no surprise that back in the days when astronomers didn't know for sure if such things as exoplanets existed, most of them considered being able to confirm their existence a pipe dream. In fact, one of the few people who believed that it was possible – and crucial enough to our understanding of the universe to be worth doing – wasn't even an astronomer. This was an engineer named Bill Borucki, who had made a name for himself in the 1960s when he designed the heatshield that was used on the Apollo spacecraft.

By the 1980s, Borucki was giving serious thought to the mechanics of the planet-hunting business, with specific reference to the transit method. As he saw it, there were basically two parts to the problem of confirming the existence of exoplanets. First, it would be necessary to design a photometer – the generic name for an instrument that measures a star's brightness – that was accurate enough to detect a dip of just a few parts per million. Second, Borucki would need to do something that had never been dreamed of before. He needed a telescope that could stare continuously at the same small part of the sky, day in and day out for years, and simultaneously monitor the brightness of at least 10,000 stars at a time. That's something that, due to the Earth's day/night and seasonal cycles, couldn't possibly be done using a ground-based telescope, so it had be located in space. And it would need to be totally unlike any other space telescope ever imagined.

Borucki arrived at this conclusion as long ago as 1984, but when he suggested it to the astronomical community, he was ostracised as a crank. Back in Chapter 1, in the context of Karl Jansky's discovery of cosmic radio emissions, we saw how astronomers can sometimes block out new ideas when they come from outsiders. To be fair to them, this may not be entirely down to knee-jerk prejudice. Astronomy, like all areas of science, has its own special language and jargon, and its own currently fashionable lines of thought. There's a risk that an outsider who speaks a different technical language, or flies off on a tangential train of thought, can sound illiterate to those on the inside.

Fortunately, it was only within the astronomical community itself that Borucki got the cold shoulder. In his day job as an engineer at NASA, his employers still thought highly enough of him to allow him to set up a small team to look in more detail at the practicalities of the method. They weren't given enough resources to make rapid progress, but by the early 1990s, they had succeeded in developing a photometer to the required specification. As for that other challenge – observing a large number of stars simultaneously – the team demonstrated its feasibility by the simple expedient of actually doing it, using a ground-based prototype.

But cracking the technical problems wasn't the end of the struggle for Borucki and his team. They submitted plan after detailed plan to NASA for a space mission they referred to as FRESIP, which stood for 'Frequency of Earth-Sized Inner Planets' – a pretty fair description of what it was designed to find out. But on each occasion, the plan was rejected – initially on the grounds that the necessary detector technology didn't exist (which it did), and later because it was deemed far too expensive (which it wasn't).

Away from NASA, however, Borucki's proposals did attract the attention of a couple of potentially important supporters, in the form of Jill Tarter and Carl Sagan. They were among the few professional astronomers who took the possibility of exoplanet detection seriously in those days. From their own efforts in that area, Tarter and Sagan knew what a hard sell the subject was and suggested to Borucki that at least part of his problem may have been the uninspiring name, FRESIP, that he'd lumbered his project with. They recommended that, as an unabashed sales gimmick, he might like to change its name to Kepler, after the 17th-century astronomer who originally worked out the laws of planetary motion. Borucki, who was attached to the name he'd originally come up with, couldn't see much point in changing it – but he deferred to their greater experience and changed it anyway.

In hindsight, now that we know everything Kepler has achieved, it's difficult to imagine it under any other name. But it's unlikely that the name change itself was what swung Borucki's project from a perpetual loser to a must-have in NASA's eyes. Something else happened around the same time, in the mid-1990s, that had a much greater impact. The first exoplanets were found using ground-based telescopes, and the subject promptly switched from a fringe fantasy to a cutting-edge scientific one. After doggedly submitting his proposal – dutifully updated each time to meet the latest round of criticisms – once every two years from 1992, Borucki's team were finally given the go-ahead in 2000.

Almost two decades after a near-crackpot concept called FRESIP had been born on paper, Kepler blasted off into space in 2009. It wasn't heading for a spot in Earth orbit, but for a much bigger orbit around the Sun. This had the advantage that it gave the telescope a completely uninterrupted view of the sky, without the Earth repeatedly getting in the way. It also meant that Kepler could remain in a fixed orientation with respect to the stars, so that its optics were always pointing at the same patch of the sky, in the constellations of Cygnus and Lyra. Throughout its life, that's all it would ever stare at, carefully monitoring those stars in the hope of spotting transit-style dips in their light curves.

Two years after Kepler's launch, in February 2011, Borucki presented a detailed analysis of its initial observations at a NASA press conference. These only covered the telescope's first four months of operation, but during that time, it had found over 1,000 'planetary candidates'. This was a cautious term the team used to describe something

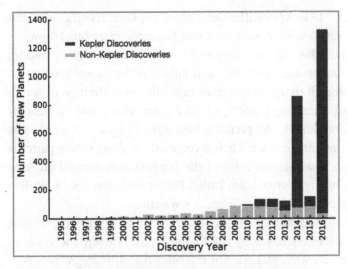

**This graph, showing the status of exoplanet discoveries seven years into the Kepler mission, shows how it came to dominate the field.**

NASA

that behaves in the way expected of a planet, but still needed formal confirmation through follow-up observations. In fact, over 90 per cent of those preliminary candidates later proved to be real planets, and many more of them were discovered as time went on.

Of course, not all of Kepler's planets were Earth-like, which was the mission's real interest – as embodied in its clunky original name, FRESIP: 'Frequency of Earth-Sized Inner Planets'. But there were notable successes in this area too, with the first genuinely Earth-like planets, Kepler-62e and Kepler-62f, being announced in April 2013. Orbiting the same host star, these both lie in its 'habitable zone' where liquid water can exist, and they're both similar in size to our own planet.

Discoveries like these are great headline-grabbers, but there's a lot of hard work that has made these breakthroughs possible. In the course of almost ten years of operation, Kepler measured the light curves of half a million stars – and all that data had to be carefully sifted through in search of transiting planets. Up to a point, this could be done by computers, but even the best algorithms aren't as good as the human brain when it comes to spotting subtle patterns in raw data. So a lot of the legwork was farmed out to a citizen science team called Planet Hunters, led by professional astronomer Tabetha Boyajian.

As it turned out, however, the most attention-grabbing discovery made by the Planet Hunters team had nothing to do with planets. On the other hand, it might just possibly have had something to do with the topic everyone's really thinking of when they talk about other planets – namely extraterrestrial life. The discovery I'm talking about relates to the concept of 'alien megastructures' – which sounds like something out of science fiction but is a theoretical possibility in science fact too. The idea is that a sufficiently advanced alien civilisation might decide that naturally formed planets – all different sizes and at different distances from their host star – are a waste of material and proceed to reshape them into more efficiently designed structures that orbit at much more convenient distances.

A few years before Kepler was launched, in 2005, the French astronomer Luc Arnold wrote a paper called 'Transit Light-Curve Signatures of Artificial Objects'. If you think about everything we've discussed in this chapter, you can probably see what he was getting at just from that title. He made the point that artificially created megastructures, when they happen to pass in front of their host star as we're observing it, will produce a distinctive light curve of their own – analogous to, but different from, that of a naturally formed planet.

The big discovery the Planet Hunters team made related to a star labelled KIC 8462852 in the Kepler input catalogue. When they analysed its light curve in 2015, they found a pattern of dimming that looked nothing like a planet, but very like what Arnold had predicted for an alien megastructure. When the discovery was announced, in a paper co-authored by Tabetha Boyajian and almost 50 other members of the Planet Hunters team, it was gleefully picked up by the mass media as 'likely' evidence of intelligent aliens. The star in question acquired the nickname 'Tabby's Star' after its lead discoverer, and space artists quickly came up with imaginative illustrations of it surrounded by swarms of enigmatically alien-looking structures.

Sadly, while alien-built structures are perfectly possible in theory, they almost certainly aren't the cause of the anomalous dips in Tabby's Star's light curve. After further analysis, it's much more likely that what's being observed is simply a large cloud of dust around the star. The clincher was the finding that the size of the dips varies with the wavelength that's being measured. As Boyajian herself put it:

> If a solid, opaque object like a megastructure was passing in front of the star, it would block out light equally at all colours. This is contrary to what we observe.

But if Tabby's Star ended up as something of a disappointment, Kepler more than lived up to expectations in its intended field: discovering exoplanets. By the end of its working life – which was extended from an originally planned 3.5 years to 9.5 – Kepler had discovered at least 2,700 new exoplanets, more than two-thirds of all those that were known at that time. When you recall that it was just looking at a single patch of the sky, and that it could only detect planets that happened to be aligned in a way that produced a transit visible from Earth, then that's a truly extraordinary number. On the basis of Kepler's results, statistical analysis suggests that at least 80 per cent of all stars – with the possible exception of the very oldest generations – have at least one planet.

While it's interesting to detect any new planet, what people really want to find – and I'm talking about astronomers themselves, as well as the general public – are habitable, Earth-like planets. Of course, using the kind of data that Kepler acquired, we can never know for sure if an exoplanet is habitable, but we can make an educated guess. By analysing the shape of its light curve during a transit, it's possible to work out a planet's size – and if it's close to the Earth's, there's a fair chance it has a similar composition too – as well as its distance from its host star. From the latter, we can estimate whether it's at the right temperature for liquid water to exist on its surface, which is generally considered the most important prerequisite for life and habitability.

By these criteria, Kepler has discovered at least 30 planets that appear to be reasonably Earth-like and potentially habitable. I've already mentioned the first two of these to be discovered, Kepler-62e and Kepler-62f, and we met another of them, K2-18b in Chapter 2 (the one where follow-up observations by Hubble showed the presence of water in its atmosphere). To pick just one other example, there's Kepler-452b, which the media sometimes refers to as 'Earth 2.0'. Not only is this planet physically similar to our own, but – unlike the other three planets I mentioned – it orbits a star that closely resembles our own Sun.

All these planets are interesting enough to warrant further investigation, but there's one big catch. Even if we were lucky enough to discover clear signs of life on them, there's nothing we could do about it because they're too far away from us. K2-18b is 120 light years from us, Kepler-62 is 980 light years away and Kepler-452b an eye-watering 1,800 light years away. If we tried sending a message to even the closest of these, it would be 240 years before we could hope to receive a reply, since any messages would have to travel at the speed of light. This clearly isn't going to make for a very coherent conversation, even if there's anyone still around to continue it.

What we really want to do is discover Earth-like planets that are close neighbours of ours, a few tens of light years away at the most. Kepler didn't find anything in this cate-gory – but then again, it wasn't specifically designed to do that. Its immediate successor, on the other hand, was.

## TESS

Launched in April 2018, NASA's Transiting Exoplanet Survey Satellite is different from Kepler in several ways. It's some-what smaller in size, belonging to the same Explorer class of spacecraft as COBE and WMAP (for purists, TESS is officially Explorer 95). In contrast to Kepler's hefty 1.4-metre-diameter primary mirror, TESS doesn't have a mirror at all. Rather than a reflecting telescope, it consists of four refractors, each with a main lens just over ten centimetres in diameter. In many ways these resemble cameras as much as telescopes.

If you're into photography jargon, then you can think of them as having lenses of 147 millimetres focal length and f/1.4 aperture.

But don't get the idea that TESS is somehow technologically substandard. Remember that its job isn't to peer into the depths of the universe or produce high-resolution images like Hubble. All it needs to do is measure the light curves of thousands of nearby stars, and that's a task it's eminently suited to. Its photometric precision – in other words, the smallest amount of dimming in a star's light curve that it can detect – is around 50 parts per million, as good as anything achieved by Kepler.

Another difference between the two missions is implicit in TESS' full name, which includes the word 'satellite'. As we saw in the previous chapter, this term strictly speaking only applies to a spacecraft that orbits the Earth – which isn't true of Kepler as it orbits the Sun slightly further out than the Earth itself. TESS, on the other hand, does orbit the Earth – but it's one of the most peculiar orbits any spacecraft has ever occupied.

The usual situation for satellites is a near-circular orbit that's relatively close to the Earth's surface. Hubble is a case in point here, orbiting at a constant altitude of 540 kilometres – which, when you're talking about outer space, is hardly any distance at all. Some satellites are located at higher altitudes than this, and the higher they are, the longer they take to complete an orbit. In the case of Hubble, that's around 95 minutes, but by the time you get up to 35,800 kilometres, it takes a whole day for a satellite in a circular orbit to go once around the Earth. Because the Earth itself is rotating at the same rate, this means the satellite remains effectively stationary relative to the Earth's surface. This 'geostationary' orbit is particularly useful for certain types of satellite – but, with a few exceptions, satellites are rarely found at any higher altitude than this. TESS is one of the exceptions.

Imagine an elongated, oval-shaped orbit that, at its closest point, is 108,000 kilometres from the Earth – three times as far as a geostationary satellite – and at its furthest point is 375,000 kilometres away, roughly the same distance as the Moon. That's the orbit that TESS follows. It has the advantage that, like Kepler's Sun-centred orbit, it puts TESS far enough away that the Earth doesn't keep blocking its view of the sky. TESS completes exactly two orbits for every one orbit of the Moon, resulting in an orbital period of 13.7 days.

TESS' four cameras – or telescopes, if you want to call them that – point at different parts of the sky. They're arranged in such a way that their fields of view – 24° compared to Kepler's 15° – have a slight overlap, so that TESS sees a long strip of sky around 90° long by 24° wide. TESS then stares fixedly at its patch of sky just as Kepler did – but whereas Kepler did that indefinitely, TESS moves on to another patch after two orbits, or about four weeks. This means that over the course of two years, it's able to observe virtually the whole sky – as the following diagram shows.

When the results of TESS' first two years of observation were released in March 2021, they included over 2,000 exoplanet candidates – a figure that more than doubled with a follow-up data release a year later. So it's clear that TESS is just as effective at finding planets as Kepler was, although most of its initial findings remained as 'candidates' rather than confirmed planets. Getting to the latter stage requires further observation, either by TESS or another telescope, to verify that what is being seen really is a genuine planet. Of those discoveries that have been confirmed, at least

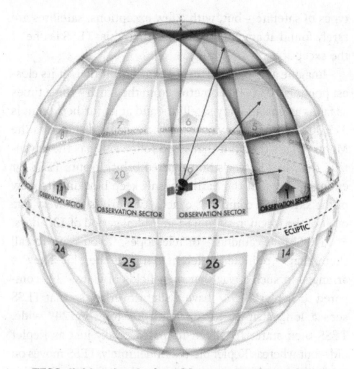

**TESS divides the sky into 26 sectors and spends two orbits – around four weeks – observing each in turn.**

NASA

twenty are Earth-sized planets within 70 light years of the Solar System, although most of those probably lie outside their star's habitable zone.

Of course, we shouldn't be so parochial as to imagine that a planet has to be closely similar to our own in order to be worth studying. Astronomer Niall Deacon has likened that attitude to British tourists arriving in an exotic far-flung destination and immediately saying, 'Let's find a fish and chip shop.' In fact, TESS – like Kepler before it – has made plenty of intriguing discoveries that are distinctly non-Earth-like in nature.

Many of TESS' earliest fully confirmed planets are in this category, for a reason that is fairly obvious if you think about it. The best way to confirm the existence of a planet is to observe multiple different transits – and since we're going to see at most one transit per orbit of the planet around its host star, that's going to happen quickest in the case of a planet with a very short orbital period. Since exoplanets follow Kepler's laws just as much as the Solar System's planets do, this means the planet needs to be located extremely close to its star – and is therefore almost certainly much too hot to be habitable.

An extreme case of this is the exoplanet Gliese 367b, which was observed by TESS between February and March 2019. During that time, it was seen to make dozens of transits, because its orbital period is an astonishingly short 7.7 hours. In other words, a 'year' on Gliese 367b lasts less than a third of an Earth day. This definitely puts it in the 'interesting' category – and fortunately, it's close enough to us, at a distance of just 31 light years, that astronomers can study it in detail. It's a small planet, somewhere between Mars and Earth in size, and probably has a similar chemical composition to the Solar System's innermost planet, Mercury. The latter has a reputation for being hot, with a daytime surface temperature up to 430° Celsius – but Gliese 367b is far hotter still, possibly reaching as high as 1,500° Celsius.

TESS has also turned up its own counterpart to the enigma of Tabby's Star, in the form of a star labelled TIC 400799224 in the TESS input catalogue. This displays a strange pattern of brightness changes that's distinctly reminiscent of Tabby's Star, characterised by sudden, very deep dips that are nothing like those expected from a planetary transit. After further investigation, this appears to be not one but two stars orbiting around each other, with one of them – rather like Tabby's Star itself – surrounded by a dense cloud of dust, or possibly the debris from a shattered asteroid.

Another anomalous light curve belongs to a star called TOI-178, where TOI stands for 'TESS Object of Interest'. The latter designation means just what it says: the object looked interesting enough on first detection to be added to a list of candidates worthy of follow-up observation by other telescopes. In this case, the original TESS data suggested there were three planets orbiting the star, in a way that made them seem strangely coordinated with each other. After further investigation, the reality turned out to be even more intriguing than this. In fact, there appear to be six planets in total, the outer five of them following mutually 'resonant' orbits – in the same way that TESS' own orbit is in resonant harmony with the Moon's. In the case of TOI-178, for every orbit of the outermost, slowest-moving planet, the next one in makes one and a third orbits, the middle planet two orbits, the next one three orbits and the fifth planet six orbits. This means that every now and then two or more of the planets can transit at the same time, resulting in the strange light curve pattern seen by TESS.

So TESS has made plenty of interesting discoveries already, but its main purpose is to act as the first stage in a longer and more important process. Its basic task here is to identify a few dozen potentially habitable planets, ideally within a few tens of light years of the Solar System, that can subsequently be targeted by other telescopes looking for signs of life. At the most obvious level, this may mean radio signals – or maybe even megastructures – that indicate the presence of advanced civilisations, but even if the planets only harbour much lowlier forms of life it might still be possible to detect it.

We saw in Chapter 2 how the Hubble telescope has already been employed to detect the presence of water in exoplanet atmospheres. The technique it uses, transmission spectroscopy, works by separating out the spectrum of a transiting planet from that of the star behind it. Water is a particularly easy substance to spot in the resulting spectrum, and it's generally considered a basic prerequisite for the presence of life – but in itself it's not a direct indicator of biological activity. On the other hand, there are other substances that might be found in a planetary atmosphere that – on Earth, anyway – are usually only created by lifeforms. These 'biosignatures', as they're called, include oxygen, ozone and methane.*

Of these, oxygen is probably the most important biosignature of all – although not for the reason that might immediately spring to mind. We naturally think of oxygen as an essential *prerequisite* for life because it's what we need in order to breathe. But the oxygen in the Earth's atmosphere is only there because it's constantly being produced by other forms of life, such as plants and bacteria. Without them, atmospheric oxygen would disappear very quickly, because it's a highly reactive substance that prefers to latch onto other molecules than to remain in its pure form. So if we see oxygen in an exoplanet's atmosphere, it's a very strong indication that there's some kind of life there.

* For a more complete discussion of biosignatures and their implications, see my book *Astrobiology*, also in the Hot Science series.

One of the originators of the transmission spectroscopy technique, Sara Seager of the Massachusetts Institute of Technology, just happens to be the deputy science director of TESS. She's also a member of the exoplanet science team on NASA's James Webb Space Telescope, the giant successor to Hubble that was launched at the end of 2021. We'll talk about it in detail in Chapter 6 – but it's safe to say that, with Seager on the team, it's going to be doing a lot of transmission spectroscopy on the most promising-looking exoplanets. And these, in turn, will come from the shortlist produced by TESS – which has been likened to a 'finder scope' for JWST.

The study of exoplanets has come a long way since Bill Borucki had such an uphill struggle back in the early 1990s to convince the astronomical community that it was something worth doing. It's become one of the hottest topics in astronomy – and, not surprisingly, there are other specialised space telescopes in the pipeline beyond TESS. One that's particularly worth mentioning is an upcoming ESA mission called PLATO – a rather contrived acronym for 'PLAnetary Transits and Oscillations of stars' – which is scheduled for launch in 2026.

Just as ESA's Planck telescope was a 'bigger and better' version of NASA's WMAP, so PLATO can be thought of as a bigger and better version of TESS. It has a similar basic design, but with a total of 26 separate cameras compared to TESS' four, and an improved photometric precision of just 30 parts per million. In common with an increasing number of space telescopes these days, it will be positioned at the L2 point rather than Earth orbit, and ESA's goal is for it to survey at least a million bright stars in search of planetary transits (PLATO also has a secondary mission to study seismic activity in stars, which is where the 'O' for oscillations in its name comes from).

Around three years after PLATO, ESA intends to launch a follow-up exoplanet mission called ARIEL (Atmospheric Remote-sensing Infrared Exoplanet Large-survey), also to be located at L2. This will use the transmission spectroscopy method to study planetary atmospheres – but unlike Hubble or JWST, it will do this full-time, rather than in between other tasks. This will allow ARIEL to cover a much larger sample of exoplanets, possibly as many as a thousand of them.

As scientific surveys go, these numbers sound impressive – with ARIEL studying a thousand different planets, and PLATO observing a million stars. But for sheer industrial-scale astronomy, these pale into insignificance compared to another ESA space telescope, which has measured the properties of almost 2 *billion* astronomical objects. It's called Gaia, and we'll meet it in the next chapter.

# MAPPING THE GALAXY 5

During the seven or so years that I worked in astronomical research, the field I was involved in related to the structure and dynamics of galaxies. Or, to put it in everyday terms, what shape they are, and how the stars inside them move about. This sounds simple enough, but if you think about it for a moment, you'll see there's a complication that means we can never study those things directly. A photograph of a galaxy only shows us its two-dimensional projection on the sky, from whatever direction we happen to be viewing it – not its actual three-dimensional shape.

There's a similar limitation when it comes to measuring the motion of stars inside a galaxy. We saw in Chapter 1 how the change in wavelength of a star's spectral lines due to the Doppler shift can tell us something about its speed relative to us. But this only gives us the component of the star's velocity along the line of sight, not its full three-dimensional motion. So the best we can do is make a set of computer models based on realistic physics, and then see which of them – in whichever viewing orientation – best matches a distant galaxy's observed appearance and spectroscopic measurements. This will give us a fair idea of what the galaxy *probably* looks like – but it's still only a semi-theoretical model, not an actual observation.

When it comes to our own galaxy, we're in a much better position than we are with any external one. For one thing, it's much easier to make out individual stars in our own galaxy without them all merging into an incoherent blur, as they tend to do in distant galaxies. We can't see all 100 billion of them, because most of them are too faint, but we can see enough of them to make up a representative sample. For the stars we can see, we're potentially able to measure their two-dimensional positions in the sky with extreme precision – a process known as 'astrometry'. And as we'll see in a moment, there are other astrometric techniques that allow us to measure a star's distance from us, and the part of its motion that lies in the plane of the sky – complementing the other part, perpendicular to this, that we can obtain via spectroscopy.

Putting all those measurements together, we can establish the full three-dimensional position and motion of a star, which is exactly what studies of galactic structure need. If there's a catch in all this, it's that doing it for anything but the very closest stars requires a much greater measurement accuracy than anything that could be obtained with a ground-based telescope. But this is a book about space telescopes, so you probably saw that coming anyway.

## Hipparcos

The first space telescope designed to do astrometry on a large scale was ESA's Hipparcos, launched in 1989. The name is a nod to the Greek astronomer Hipparchus, who lived in the 2nd century BCE and was one of the earliest pioneers of astrometry, as well as another of ESA's carefully contrived acronyms – this time standing for 'HIgh Precision PARallax COllecting Satellite'.

'Parallax' is the key word here, and it is one that many people will be familiar with – although not necessarily in an astronomical context. Parallax is the optical effect that makes nearby objects appear to change position more noticeably than distant objects when the viewpoint is changed. Anyone who's watched an old cartoon, or played a 2D video game, will know how they use 'parallax scrolling' to exploit this effect. As the viewpoint moves from, say, left to right, the background is presented as two or three layers that move from right to left at different speeds, thus giving an illusion of depth.

The cartoon or game animators know the distance they want an object to be and use it to work out the apparent speed it has to move in order to give the correct parallax effect. On the other hand, when astronomers exploit parallax, they use it the other way around. The unknown in this case is the distance to a star, but if they're able to measure its apparent motion as the viewpoint changes, they can then use parallax to estimate how far away it is.

There's a simple way to understand what I'm talking about that you can try right now.* Look out of the window and hold a pencil steadily in front of you at arm's length. If you alternately close your left and right eyes, you'll see how the pencil appears to jump in the opposite direction – right

---

* I'm assuming you're sitting comfortably at home. If you happen to be on a train, bus or plane, you'll have to do the experiment in your head.

and left – relative to the view outside. That's a consequence of parallax, because the angular direction to the pencil is slightly different when seen by your left and right eyes. I'm not expecting you to go to all the trouble of measuring the precise angular difference, but if you did – and if you knew the exact distance between your eyes – then you could use a simple geometrical formula to work out the distance to the pencil.

When it comes to measuring the distance to stars, there's a natural counterpart to the separation between our eyes that astronomers can make use of. This is the difference in the position of the Earth as it orbits around the Sun on two occasions six months apart. Because the Earth–Sun distance is defined to be one 'astronomical unit', or 1AU, then the separation between the Earth's viewpoint in, say, January and July is 2AU.

The catch, of course, is that the stars are so far away that – even with a baseline of 2AU – the angular changes involved are incredibly tiny. As I mentioned back in Chapter 1, astronomers measure small angles using a unit called an 'arcsecond', with the same number of arcseconds making up a degree as there are seconds in an hour – 3,600.

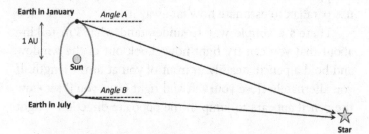

**The parallax effect allows the distance to a star to be calculated from the difference in its angular position as measured on two occasions six months apart.**

Even the very closest star beyond the Sun has a parallax of less than one arcsecond.

At this point, it's worth a brief aside to clear up a minor mystery that may have baffled anyone who has heard professional astronomers talking about their subject. When discussing cosmic distances, astronomers almost always do so in terms of 'parsecs' rather than light years. The latter, which is just the distance that light travels in a year, is a simple enough concept that anyone can understand. A parsec, on the other hand, is the distance at which an object would have a parallax of one arcsecond – which is pretty obscure even if you know what 'parallax' and 'arcsecond' mean.

So why use parsecs? Your guess is as good as mine. It's not because it's the older of the two terms – 'light year' was well established long before 'parsec' was coined in 1913 – nor because it's a more convenient size, being just 3.26 times bigger than a light year. Whatever the justification for it, as a science populariser, parsecs drive me mad. Why do scientists have to use jargon that seems deliberately designed to confuse outsiders? Anyway, I have decided to use only light years in this book.

Getting back to the measurement of stellar distances using parallax, there's a second problem besides the smallness of the angles involved. When I described the pencil analogy earlier, I was careful to say that you need to hold the pencil 'steadily'. If it was moving while you blinked first one eye and then the other, it would be much harder to work out its distance from the change in angle. Yet that's exactly what happens in the case of stars because they don't stand still. Each star has an intrinsic velocity due to its orbital motion within the galaxy – and, as we saw earlier, determining this velocity is another key objective of astrometry.

Although the effect is much too small – and slow – to be detected without specialised instruments, a nearby star follows a wiggly path across the sky relative to more distant stars. This is made up of two separate motions. Firstly, there's a steady drift in a fixed direction – referred to as the star's 'proper motion' – due to its intrinsic velocity relative to the Solar System. Secondly, there's a yearly oscillation superimposed on this due to the parallax effect caused by the Earth's motion around the Sun. We're interested in measuring both these motions, but to disentangle them we need to make a whole series of measurements spread over several years to work out which bit of a star's apparent motion is real (its proper motion) and which is an illusion caused by our own movement (its parallax).

Now that we've seen how astrometry boils down to measuring very small angular distances, it's more obvious why ground-based telescopes can't do it for anything but the very closest stars. Images taken through the Earth's atmosphere are simply too fuzzy to make measurements with the necessary accuracy. When Hipparcos was launched in 1989, only about 8,000 stars had known parallaxes. By the end of its mission in 1993, that figure had been multiplied by fifteen to almost 120,000.

Of all the space telescopes discussed in this book, Hipparcos is the one that came closest to failure even before it commenced operation. The original plan called for a very high Earth orbit – at that special 'geostationary' altitude of 35,800 kilometres. The standard way of reaching this is via an intermediate 'geostationary transfer orbit', which is a highly elongated ellipse – the upper end at the desired final altitude and the lower end much closer to Earth. In the case of Hipparcos, this lower altitude was just 500 kilometres, comparable to that of the Hubble telescope.

The next step – when things are going according to plan – is to fire an upper-stage rocket to boost the satellite into a circular orbit at the geostationary altitude. But on this particular mission – to the horror of the watching scientists – the rocket failed to fire, leaving Hipparcos stuck in that rather inelegantly lopsided transfer orbit. This was a potential disaster because it meant that each and every orbit took the satellite through the high-energy radiation belts that surround the Earth – a region that most space missions try to spend as little time in as possible. Fortunately, though, Hipparcos proved hardier than expected, and it continued operating even longer than had originally been planned despite being exposed to this unanticipated hazard.

The measurements of stellar position and parallax that Hipparcos made were accurate to within about a thousandth of an arcsecond. A star with a parallax of that amount lies, by definition, at a distance of 1,000 parsecs – or, since I promised not to use that measurement, 3,260 light years. But if Hipparcos measured a parallax that small, you could be forgiven for taking it with a very large pinch of salt, since it's right at the limit of the telescope's accuracy. Measurements are really only going to be trustworthy if they're ten or more times the limit of precision, which would allow you to say that they're accurate to plus or minus 10 per cent or better. This puts Hipparcos' effective range at around 326 light years – and indeed, most of the 120,000 stars it measured were within this distance.

That's certainly impressive, but in terms of 'mapping the galaxy' – which is what this chapter is about – is it enough? Remember that, as I said in Chapter 1, the centre of the galaxy is about 30,000 light years away, so Hipparcos' data doesn't even get close to it. To be honest, the question hardly matters any more, because – as we'll see shortly – the successor to Hipparcos expanded our horizons by a factor of 100. But even with relatively short-range measurements, it's possible to make deductions about the large-scale structure of the galaxy if they're suitably combined with theoretical models – and since it's a topic I was involved in myself, I'm going to tell you about it.

Back in the mid-1980s, I spent a couple of years working as a postdoctoral research assistant to James Binney – a top-flight theoretician who literally wrote the book on galactic dynamics.* I'm not sure if the idea of using local observations to determine global structure was original to him, but I'm pretty sure that a paper I co-wrote with him was the first to discuss it in detail. It was called 'Solar Neighbourhood Observations and the Structure of the Galaxy', and it appeared in the August 1986 issue of *Monthly Notices of the Royal Astronomical Society*. Here's the beginning of it – which, apart from a few pieces of technical jargon that I'll explain in a moment, I think describes the concept pretty well:

> If we could travel around the galaxy as easily as we can travel around the Earth, the study of galactic structure would be a very simple matter. We could build up a detailed picture of the phase-space distribution of the galaxy by direct six-dimensional cartography, without needing to know anything about stellar dynamics. Unfortunately, however, we are confined to the Solar System, and we can

---

* Well, co-wrote, anyway: James Binney and Scott Tremaine, *Galactic Dynamics* (New Jersey: Princeton University Press, second edition, 2008).

only observe the galaxy from this one particular viewpoint. Nevertheless, with the advent of Hipparcos and the Space Telescope, there should soon be a very large body of observational data available on stars in the immediate neighbourhood of the Sun. To what extent do such solar-neighbourhood observations contain information on the large-scale structure of the galaxy, and what are the most efficient methods of extracting this information?

The great advantage that a galactic astronomer has over any other would-be armchair cartographer is Jeans' theorem. This states that, for a well-mixed stellar population, the phase-space density can be taken to be a function only of the isolating integrals of motion. Thus, we should picture the galaxy not as a swarm of moving points in real space, but as a set of fixed points in isolating-integral space.

As spookily sci-fi as it sounds, 'phase space' is just a convenient way of referring to the six 'dimensions' that characterise a star's position and motion: its coordinates in real three-dimensional space and the corresponding three components of its velocity. Jeans' theorem, named after astrophysicist and science populariser Sir James Jeans, provides a way to convert what may look like a very complicated distribution of stars in this six-dimensional space into a smaller number of dimensions based on physical properties such as energy and angular momentum (these are the 'isolating integrals' referred to above).

The point (now that I'm finally getting to it) is that the sample of nearby stars measured by Hipparcos actually encompasses a wide range of energies and angular momenta. So when viewed through the lens of Jeans' theorem, it isn't really a 'local' sample at all – it's potentially much more representative of the galaxy as a whole than it appears at first sight. This makes the data much more useful in choosing between computer-generated models of large-scale galactic structures.

Anyway, that's all history now – getting on for 40 years ago. Astronomers like James Binney are still building galaxy models and fitting them to observational data, but these days the data comes from Hipparcos' successor, Gaia – and it's far from being limited to the solar neighbourhood.

## Gaia

The name Gaia started out as another acronym, standing for 'Global Astrometric Interferometer for Astrophysics'. An interferometer is a special kind of telescope, and one that's quite complicated to describe. Fortunately, I don't have to because, after ESA chose this name for their follow-up to Hipparcos, the telescope's design changed and it's no longer an interferometer. But they liked the name so much they stuck with it.*

In the event, Gaia is really just a bigger and better version of Hipparcos. Much better, in fact; as I said earlier, its angular precision is improved by a factor of 100. While Hipparcos could measure angles as small as a 1,000th of an arcsecond, Gaia can get down to a 100,000th of an arcsecond – or, if it's easier to think of it that way, ten micro-arcseconds. Back

---

* As a purely subjective view, I have to admit I prefer the names of ESA missions to NASA's. To me, ARIEL, PLATO, Hipparcos and Gaia have an impressive gravitas that WMAP, TESS, Hubble and Webb don't. But maybe that's because I'm a stuffy old European rooted in the culture of the past.

in Chapter 1, I mentioned that an arcsecond is the angular size of a five-pence coin, or an American dime, seen at a distance of four kilometres. But Gaia can do 100,000 times better than that – more like spotting a coin on the surface of the Moon.

ESA's plan, which was truly grandiose in scale, was for Gaia to measure the parallax of at least a billion stars. That's 8,000 times as many as Hipparcos – around 1 per cent of all the stars in the galaxy. Most importantly of all, it's an evenly distributed sample across the whole galaxy, not one that's restricted to the 'solar neighbourhood', as was the case with Hipparcos.

There was a twenty-year gap between the end of the Hipparcos mission in 1993 and the launch of Gaia in December 2013. Unlike its predecessor, Gaia wasn't heading into Earth orbit, but to L2 – the second Lagrange point – like the WMAP and Planck telescopes discussed in Chapter 3. As we discovered, L2 has become an increasingly popular location for space-based observatories, combining the uninterrupted views and ultra-cold temperatures of deep space with the benefit of a highly stable orbit that's relatively close to Earth.

From this location, Gaia views the galaxy using an unusual optical arrangement that was pioneered by Hipparcos. Instead of a single telescope, it has two identical instruments pointing in different directions, at a fixed angle of 106.5° apart. It then combines the two resulting images and processes them together, in such a way that it only has to measure the relative – as opposed to absolute – angular positions of stars. This may sound unnecessarily complicated, but the upshot is that it's much easier for the telescope to measure the incredibly tiny parallax angles it needs to do its job.

**The Gaia telescope looks in two directions at once, as illustrated in this still image from an ESA video.**

ESA

From the outside, with its cylindrical body, Gaia looks rather like other space telescopes such as Hubble or Kepler. But that's a misleading impression since its twin optics actually look out of the sides of the cylinder instead of the end. There's also another fundamental difference from those Earth-orbiting telescopes in that Gaia is designed with a permanently 'hot side' and a permanently 'cold side'. This is possible because of that special location at L2, which means that the two strongest sources of heat – the Sun and the Earth – are always in the same direction. The spacecraft needs to have a good view of both of them, the Sun to provide energy for its solar panels, and the Earth to send back all the data it's collected. But the telescope itself – the part that's looking into deep space – needs to be kept as cold as possible for its electronic components to function properly.

So Gaia's optical system and scientific instruments are mounted on the side away from the Sun, protected from its heat by a ten-metre-diameter sunshield. Designed to fold out from the spacecraft after launch, this keeps its cold side at the decidedly chilly temperature of –110° Celsius, which is what the instruments need in order to achieve the desired measurement accuracy.

On the hot side of the sunshield, two rings of solar panels constantly face the Sun to generate the power – around two kilowatts – needed to operate the spacecraft's electrical systems. Also located on this hot side is Gaia's all-important downlink antenna, through which it transmits the data it's acquired back to ESA's well-established net-work of radio-telescope-like ground stations. And we really are talking about a lot of data here; over the entirety of its mission, Gaia is going to collect hundreds of thousands of gigabytes of data.

All this data is sent back to the ground stations on Earth at a rate of up to five megabits per second for eight hours a day. 'So what?' you may reply, 'that's not a massive amount of data by today's standards.' But there's some-thing that makes this data transfer a much bigger headache for the engineers than it might otherwise be. If they designed the spacecraft with an ordinary, mechanically steered radio dish, the vibrations caused by its motion would scramble all the ultra-precise astronomical measurements the telescope is trying to make. So Gaia has to use a special, high-tech kind of antenna – technically known as a 'phased array' – that can steer its beam electronically, without the need for any moving parts.

Gaia arrived at L2 a few weeks after launch, and then spent several months testing and calibrating its complex suite of instruments before starting serious observations in July 2014. The mission was originally planned to last five years, but the telescope is performing so well – and living up to all of ESA's ambitious expectations – that this has been extended on more than one occasion. At the time of writing (mid-2023) it's still going strong.

I said earlier that Gaia's primary purpose is to measure the parallax of at least a billion stars. That's a tremendously difficult challenge, particularly when you think that getting an accurate value for parallax – and disentangling it from the star's proper motion, which we also want to know – isn't something that can be done with just one or two measurements. In fact, Gaia is going to have to look at every one of those billion stars up to 70 times each. And even that isn't the end of the story.

Something that I haven't explicitly mentioned yet is that Gaia doesn't just make astrometric-type measurements. It also has a spectrometer on board that allows is to determine chemical compositions and radial velocities – in other words, the motion of a star directly towards or away from us, as distinct from its proper motion in the plane of the sky – for a selected subset of the stars it observes.

All that raw data, after it's been transmitted down to Earth, is passed on to Gaia's Data Processing and Analysis Consortium, a team of 450 scientists spread across Europe and beyond. Their job – and it's one of the biggest and most challenging in the history of data processing – is to make scientific sense of all that information. The results, which are being released into the public domain as and when they become available, will amount to the first truly representative census of the galaxy's stellar population ever made.

Looking at Gaia's third data release, in June 2022, the statistics are truly mind-blowing. Containing ten terabytes of information and based on some 40 billion individual observations, the database includes:*

- Accurate positions (two-dimensional location in the sky) for 1,800,000,000 stars
- Brightness measurements for 1,800,000,000 stars
- Colours for 1,500,000,000 stars
- Parallax (distance) and proper motion for 1,500,000,000 stars
- Spectral measurements for 220,000,000 stars
- Radial velocities for 33,000,000 stars
- Variability analysis for 10,000,000 stars
- Rotational speeds for 3,500,000 stars

Not surprisingly, this is by far the largest astronomical catalogue ever published. If you'd described it to me back in 1986, when I co-wrote that 'solar neighbourhood' paper, I would have dismissed it as pure fantasy. Even today, I have difficulty getting my head round the idea of measuring parallaxes and proper motions for stars that are up to 30,000 light years away.

You'll have gathered by now that I'm a big fan of Gaia – in fact, it's possibly my favourite space telescope in this book. That's partly because it's focused on an area of astronomy that I'm especially familiar with, but also because I feel it's constantly underrated by the media – particularly in comparison with Hubble. The truth is, in terms of both engineering sophistication and scientific contribution, the two telescopes are very much in the same league. By at least one standard

---

* To avoid looking too pedantic, I've rounded all these numbers to two significant figures.

academic criterion – the number of research papers that draw on its data – Gaia is actually ahead of Hubble, with something like 1,600 papers per year directly based on its data.

Although its primary purpose is to aid astronomers in creating a map of the stars in our galaxy, Gaia itself doesn't necessarily know or care whether the object it's looking at is a star or not. So almost by accident, it's collected data on a whole host of other celestial bodies as well. On a much smaller scale than the galaxy, within our own Solar System, it's derived orbital parameters for around 150,000 objects, most of them relatively small asteroids. And on a much bigger scale than the galaxy, it's obtained brightness and spectroscopic measurements for millions of galaxies and quasars.

Between those two extremes, and perhaps most surprising of all, Gaia has actually managed to measure proper motions inside another galaxy – albeit one that's very close to our own. Bearing in mind that 'proper motion' means actually seeing a star change position due to its own intrinsic velocity relative to us as we observe it, it's astounding to think we can do that for something that's so far away as to be in another galaxy. In this particular case, it's a dwarf elliptical galaxy called Sculptor that orbits around our own Milky Way like a satellite – currently located around 270,000 light years from us. Thanks to Gaia's measurements of proper motion, combined with other observations made by Hubble, we now have a pretty accurate knowledge of the shape and speed of Sculptor's orbit.

In a more indirect way, Gaia has even contributed to our understanding of cosmology, which deals with the large-scale evolution of the universe as a whole. This came about through another collaboration between Gaia and Hubble, and it relates to something named after the latter's human namesake, Edwin Hubble. This is the 'Hubble constant' – a number that describes the rate at which the universe is expanding. Edwin Hubble first estimated this by observing a particular type of pulsating star, called a Cepheid, in other galaxies.

To whiz through the theory at lightning speed, it goes something like this. The rate at which a Cepheid pulsates depends on its intrinsic brightness. On the other hand, its apparent brightness, as seen from Earth, depends both on its intrinsic brightness and its distance. But if we measure both its pulsation rate and its apparent brightness, we can eliminate the middle-man – intrinsic brightness – and deduce the Cepheid's distance. In the case of a Cepheid located in a remote galaxy, much too far away for us to measure its parallax, this provides the best way to estimate the distance to that galaxy. At the same time, we can use spectroscopic methods to estimate the galaxy's recession speed. If we do this for a large sample of galaxies, and then plot a graph of recession speed versus distance, we can deduce the rate at which the universe is expanding – in other words, the Hubble constant.

That's the technique that Edwin Hubble employed, and it's still the best one to use today, with the greatly improved data made possible by the Hubble telescope. But there's a potential weak link in the process. The whole thing hinges on having a precisely calibrated understanding of how Cepheid pulsation rate depends on intrinsic brightness. But we only know the latter in the case of Cepheids that have accurately known distances, obtained via parallax measurements. In the days when that meant just a handful of nearby stars, the resulting calibration was approximate at best.

It's at this point that Gaia comes to the rescue. Among all its other data, it has derived accurate parallax-based distances for 15,000 Cepheids scattered throughout our own galaxy, so we now have a much better understanding of the pulsation–brightness relationship. Passing this information on to the Hubble team at STScI – led by Nobel Prize-winner Adam Reiss, who we met in Chapter 2 – has resulted in the most accurate determination of the Hubble constant to date.

So from the scale of the universe right down to asteroids in our own Solar System, Gaia has contributed to many more areas of astronomy than the galactic studies it was primarily designed for. And in February 2022, Gaia snapped an image of a brand-new object in the sky. This one was very close indeed – another spacecraft from Earth, newly arrived at the L2 point. It's called the James Webb Space Telescope, and if Gaia and Hubble can be said to be roughly on a par in terms of technological sophistication and scientific potential, then this new telescope is set to tower above both of them. We'll find out all about it in the next chapter.

# WEBB 6

Not everyone has heard of Gaia – or of COBE, Planck, Kepler or TESS, for that matter. Anyone with a passing interest in space or astronomy will be familiar with at least some of those names, but this is not the case for the public at large. That's what makes Hubble so truly remarkable because it's a name that most people *have* heard of.

Before I started on this chapter, I took a short break to write a magazine piece about the Sun, and while researching it, I came across an online comment by someone referring to 'this Hubble photograph of the Sun'. But Hubble is designed to look at incredibly faint objects, not intensely bright ones. If its operators were unwise enough to point it directly at the Sun, they would instantly burn out all its super-sensitive detectors. In fact, the picture in question was taken by a completely different NASA satellite, but the anecdote highlights the extent to which Hubble has entered everyone's consciousness. For many people, if it's a photograph of something in space, then it must have been taken by Hubble.

Needless to say, this makes Hubble a hard act to follow. It remains to be seen if NASA's successor to Hubble, the James Webb Space Telescope, will become a household name in the same way. At the time I started writing this book, it was sufficiently unfamiliar that the mainstream media still tended to refer to it by its full name, while more technical accounts used the abbreviation JWST. But will there ever come a time when the one-word name 'Webb' has the same worldwide recognition that 'Hubble' does?

NASA certainly wants this to happen, and STScI's PR team hit the ground running with the first release of Hubble-style images from JWST in July 2022, soon after it began operation. A few months later, they followed these up with the new telescope's view of Hubble's most iconic photo, the 'Pillars of Creation'.

Unfortunately, there's a downside to the Webb name that's irritating to many astronomers, and so repellent to others that they're reluctant to use it. Unlike Edwin Hubble,

**Hubble's famous photograph of the 'Pillars of Creation' (left), alongside the same view as seen by the James Webb Space Telescope.**

NASA

James Webb wasn't an astronomer but a career bureaucrat. In the 1950s, he worked in the US government's State Department and then went on to became head of NASA during the critical period of planning for Project Apollo in the 1960s. To many people in NASA, this latter part of his CV makes him a hero – but not everyone agrees. The fact is, James Webb had a dark side.

Many people will have heard of Senator Joseph McCarthy's 'witch-hunt' persecution of communists and other left-wingers during a period in the 1950s known as the 'Red Scare'. Less well known is the fact that there was a similar campaign of persecution, known as the 'Lavender Scare', against the LGBT community – and sad to say, James Webb played a major role in it. Even after he moved to NASA, Webb fired at least one employee for 'immoral, indecent and disgraceful conduct'. With a history like that, it's not surprising that many astronomers are unhappy about using a telescope named after him, and thousands of them signed a petition calling on NASA to rename the mission. Unfortunately, this didn't happen, so we're stuck with the fact that what may turn out to be the greatest telescope in history has a name that makes many people shudder.

Its name apart, there's every reason to expect great things from JWST. It's the largest telescope that's ever been put into space, comparable in mass to a large bus. From both an organisational and an engineering point of view, it's one of the most complex and ambitious missions that NASA has ever undertaken, right up there with the Apollo Moon landings. And at a cost of $10 billion, it's the most expensive scientific instrument ever built. For comparison, Hubble cost just $1.5 billion, while even CERN's sprawling Large Hadron Collider was less than half JWST's cost.

So where did all that $10 billion go? For one thing, it wasn't spent in a flash – it was spread over 24 years, during which it paid for a whole host of design, manufacturing and testing activities involving thousands of people. It is estimated that getting the telescope all the way from conception to launch took around 100 million person-hours. If you divide that into $10 billion, you'll see that each of those person-hours cost an average of $100. That may sound a lot if you think of it purely as wages, but you have to remember the money also paid for things like fuel, rent, utilities and insurance as well. No one got stinking-rich out of it, and virtually all the expenditure will have found its way back into the economy. Far from 'wasting' money, massive science projects like JWST simply recirculate it – just like any other business enterprise.

While the James Webb telescope very much follows in Hubble's footsteps as regards its scientific goals, it's far from being a straight replacement in terms of the technology it employs. Most importantly, it's designed to view the universe in a different part of the electromagnetic spectrum. We covered the latter topic in Chapter 1, but it may be worth a quick recap of the essentials here.

It's natural to think of the universe primarily in terms of its appearance in visible light – the kind that our eyes can see. But light is just one form of electromagnetic waves, which cover a much broader spectrum than the small band we're directly aware of. These waves are generally emitted by material bodies, with cooler bodies producing longer wavelengths and hotter ones producing shorter wavelengths. The longest wavelength that our eyes are sensitive to, which we perceive as the colour red, is around 750 nanometres (nm), and the shortest – violet – around 380 nm.

The reason our eyes have evolved to favour these wavelengths is that they account for the bulk of the sunligth reaching the Earth's surface. Longer wavelengths, referred to as infrared, and shorter, ultraviolet ones, find it harder to get through the Earth's atmosphere. On top of that, the Sun's peak output happens to lie bang in the visible band – so all in all, evolution got it right (no surprises there).

Given all those colourful photographs that Hubble is famous for, you might guess that it sees the same wavelengths as we do. And that's true, it does – but it sees other wavelengths too, well into the invisible-to-us bands that lie on either side of the visible spectrum. Hubble's sensors actually span all the way from 100 nm in the ultraviolet to 2,500 nm in the infrared. Webb, on the other hand, is first and foremost an infrared telescope, covering a wide range of wavelengths from 600 nm all the way to 28,000 nm. This range starts in the orange part of the visible spectrum – so Webb is unable to able to see green, blue or violet light – and then extends much further into the infrared than Hubble.

The obvious question at this point is 'why?' Why ignore a large chunk of the visible spectrum and focus on wavelengths we can't even see? To answer that, we need to take a step back and look in more detail at why astronomers get so excited about the infrared.

## Infrared astronomy

Infrared is sometimes referred to as 'heat radiation', and in terms of the way humans use and perceive it, that's not a bad description. Heat-seeking missiles and thermal imaging cameras use infrared sensors to detect heat sources, for example. But strictly speaking, 'heat' is a property of matter, caused by random motions of its constituent molecules. Hot objects emit radiation depending on their temperature, and over a wide range of what we might think of as 'normal temperatures' this radiation falls in the infrared part of the spectrum. But very hot objects can produce visible or even shorter wavelengths – for example, an object heated to 4,200° Celsius glowing red and one at 7,200° C glowing blue.

The connection between infrared and astronomy goes back a long way to its original discovery in the year 1800. This was made by William Herschel, who is best known as the discoverer of the planet Uranus. In an experiment with a prism, he split sunlight into its constituent colours and tried measuring their temperature with a thermometer. To his surprise, he found that it registered its highest temperature just beyond the red end of the spectrum, where no visible light was showing. He'd discovered what is now referred to as the 'near-infrared' part of the spectrum – 'near' simply meaning that it's located near to the visible band. The 'mid-infrared' and 'far-infrared' bands lie beyond this at progressively longer wavelengths.

Depending on their temperature, astronomical objects produce emissions across virtually the whole of the electromagnetic spectrum. Our own Sun happens to peak in the visible band, but cooler objects will be more prominent in the infrared – and this fact alone makes it interesting to astronomers. But there's a second reason on top of that. Some parts of the galaxy, such as its central nucleus and the star-forming regions in its spiral arms, are shrouded in clouds of fine dust that tend to absorb visible light. As a result, even ordinary stars like the Sun can be impossible to see at visual wavelengths if they lie behind a dust cloud.

On the other hand, infrared emissions are much less affected by dust, so they can pass through such clouds almost as if they weren't there.

This may all sound very specialised, but the fact is that those two things – being able to see cool objects and being able to look through dust clouds – hold the key to answering many of the most pressing questions in astronomy today. Those dust-shrouded regions, for example, are of crucial importance in understanding the nature of the galaxy around us. Its centre, inhabited by a supermassive black hole surrounded by a dense concentration of stars, is arguably its single most interesting feature – yet we can't see it at visible wavelengths due to all the dust that's in the way. A similar problem arises in the case of star formation; we can't watch it happening in visible light because, once again, it's obscured by clouds of dust.

Even when we are able to see a star in the process of formation, the 'protostar' stage – before the onset of nuclear fusion turns it into a proper star – is so cool that it emits most of its radiation in the infrared range. The same is true of stars that are so small nuclear fusion never really gets going in them. The study of these failed stars – or 'brown dwarfs', as they're called – is relatively new in astronomy for the simple reason that in the days before infrared telescopes, no one had ever seen one.

In writing the last couple of paragraphs, I've tried to avoid using the term 'hot topic' because, after all, we're talking about objects that aren't very hot at all – that's why we need infrared telescopes to see them. But mixed metaphors notwithstanding, I'm going to use the phrase now. Possibly the hottest astronomical topic of all these days is the search for exoplanets, but this rarely involves actually seeing the planet itself. They're almost always detected by indirect means, such as their effect on their host star's light curve during a transit.

Yet we can easily see planets in our own Solar System, such as Venus or Jupiter, by the light they reflect from the Sun – in fact, they're among the brightest objects in the night sky. So why can't we see exoplanets in the same way? The reason is that any light they reflect from their host star is completely swamped by the direct light of the star itself, which may be a billion times brighter. At infrared wavelengths, however, this discrepancy can reduce by a factor of a thousand, so the planet is only a million times dimmer than its host star. That still sounds pretty faint, but it's just bright enough that under the right conditions – and with a suitably sensitive infrared telescope – the planet can be imaged directly.

When you think of the achievements that Hubble is famous for – I mean its cutting-edge scientific achievements, not the pretty pictures that are publicly released – they span all the areas I've just mentioned: star formation, galactic nuclei, brown dwarfs, exoplanets. But there's another achievement that's possibly even more important than any of those: its 'deep field' images of very distant galaxies, some of them so far away that we see them as they were not long after the birth of the universe. This, too, is an area where infrared observations become far more useful than visible light.

It's not too difficult to see why this is. If you cast your mind back to Chapter 3, you'll remember how the cosmic microwave background, which we detect at the relatively long wavelengths of a few millimetres, actually started its journey at much shorter wavelengths some 13 billion years ago.

What's happened in the intervening time is that the expansion of the universe has stretched out the wavelength by a factor of a thousand or so. In the same way, the intense ultraviolet radiation produced by very early galaxies has been stretched out to the point where we're going to see it best at infrared wavelengths.*

As important as it is, there are two separate effects that make infrared astronomy very difficult to do from the Earth's surface. The first is due to our atmosphere, which tends to scatter, absorb and reradiate whatever infrared waves land on it. The second problem is that the Earth has countless heat sources of its own, both natural and human-made, which are likely to swamp the faint extraterrestrial sources we're interested in. So all in all, putting your telescope in space is the best way to go if you want to do infrared astronomy.

The first serious attempt to do this involved the aptly named Infrared Astronomical Satellite (IRAS), launched in 1982. I'd always thought that IRAS was a collaboration between NASA and ESA, but during the research for this book, I discovered that's not strictly true. Rather than the whole ESA consortium, the European side of the project only involved two of its constituent nations, the United Kingdom and the Netherlands. The spacecraft itself was the responsibility of the Netherlands, with the exception of the optical payload, which was supplied by the United States, along with its launch into space; after that the satellite was operated from a ground control centre in the UK.

---

* Another way to think of this (which is mathematically equivalent) is in terms of the Doppler shift caused by the high recession speeds of those distant galaxies. This causes their light to be 'red-shifted' by a factor of a thousand – actually pushing it way beyond red into the infrared.

IRAS' basic mission was to carry out the first complete infrared survey of the whole sky at a number of different wavelengths in both the mid- and far-infrared bands. This resulted in a detailed catalogue of 350,000 individual sources, ranging from small asteroids within the Solar System to stars and galaxies. Many of these sources are extremely faint, and in order to detect them, a telescope has to be cooled to just a few degrees above absolute zero. In the case of IRAS, the telescope was small enough – with a mirror diameter of 57 centimetres – that the whole assembly could be enclosed in a bath of liquid helium. As the helium evaporated, it provided the cooling needed to keep the optics at the required temperature – but it meant that once the helium ran out, the telescope ceased to be effective. That's why spacecraft that are cooled in this way have limited lifetimes – in the case of IRAS, it was only about ten months.

Between IRAS and Webb, infrared telescopes went through several generations of steadily improving capability. However, this book is not a 'Spotter's Guide to Astronomical Satellites', so I won't dive down into any great detail about these intermediate steps. I'll just briefly mention a couple of telescopes that have already been name-dropped in earlier chapters.

The first of these is the Spitzer Space Telescope, named after Lyman Spitzer, who – as we saw in the first two chapters – enumerated the benefits of putting a telescope in space as long ago as the 1940s and went on to play a role in the early development of Hubble. Like Hubble, the Spitzer telescope was originally planned as one of NASA's four 'great observatories', destined to be launched by the Space Shuttle (we'll meet the remaining two in the next chapter). But the plan had to be changed when NASA scaled back Shuttle operations after the Challenger crash in 1986.

In the end, Spitzer was the last of the four observatories to be launched – more than ten years late, in 2003 – and the only one not to be carried into space by a Shuttle. It was also by far the smallest of the four, with a mass of 850 kilograms compared to Hubble's eleven tonnes, and a main mirror of just 85 centimetres diameter. That hardly warrants the adjective 'great', or – in my opinion – the name of such a key person in the history of space telescopes as Lyman Spitzer. Why not hold the name back until it could be applied to a piece of kit that really deserved it, such as the telescope NASA lumbered with the politically controversial name of James Webb?

As it is, Spitzer's highest-profile role has been to supplement some of Hubble's images – taken in the visual and near-infrared bands – with its own observations in the mid- and far-infrared. This brings out details that would be totally invisible to our eyes, either because they're produced at wavelengths we can't see, or because they're masked at those wavelengths by dust.

Prior to JWST, the largest infrared space telescope – closer, perhaps, to Spitzer's original spec than the cut-down version NASA ended up with – originated on the other side of the Atlantic with ESA. This was the Herschel Space Observatory, named after the discoverer of infrared radiation. The Herschel telescope made a brief appearance in Chapter 3, as a joint payload on the same Ariane 5 launch as ESA's Planck spacecraft. Planck was essentially a microwave telescope – not surprisingly, given that it was designed to study the cosmic microwave background – and covered wavelengths from eleven down to 0.3 millimetres. But the short-wave end of this range lies in the grey area between microwaves and the far-infrared band – and the latter was Herschel's region of interest. Its coverage, from around 0.05 to 0.5 millimetres (or, if you like, 50,000 to 500,000 nanometres), actually overlapped that of Planck.

Like IRAS and Spitzer before it, Herschel only had a limited lifetime before its liquid helium coolant ran out. In Herschel's case, it carried enough coolant to last four years, from its launch in 2009 to the end of the mission in 2013. But it still holds a record that may come as a surprise to some people: it had the biggest single telescope mirror ever launched into space. At an amazing 3.5 metres in diameter, it was virtually half again as big as Hubble's. And for the purely practical reason of having to squeeze everything inside a launch vehicle, it's a record that's unlikely to be surpassed in future. If you want a space telescope that's bigger than Herschel, your best option is to use a mirror that's made of several smaller segments that can be folded up for launch. That's exactly what NASA did with JWST.

Webb's main mirror is made up of eighteen hexagonal segments, each 1.3 metres across. NASA did consider

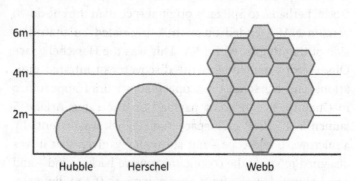

**ESA's Herschel telescope had a bigger mirror than Hubble – but Webb, with its segmented mirror, is even bigger.**

making these from glass, like Hubble's mirror, but settled on a metal called beryllium. This has the combined advantages of being both strong and light, and it's not uncommon to find it in high-end aerospace applications. As with Hubble's glass mirror, the beryllium had to be shaped to extremely high precision in order to focus images with the required clarity. And unlike the case with Hubble, Webb's technicians got the polishing right first time. Once it had been done, the mirror segments were coated with an ultra-thin layer of pure gold, to maximise their reflectivity at infrared wavelengths.

When the eighteen segments are put together, they result in an effective diameter of 6.5 metres for the main mirror. That's around 2.7 times as big as Hubble's mirror, which means that Webb will be able to see even fainter objects. The larger aperture allows more photons to reach the mirror from a distant source, before being focused down onto Webb's array of detectors. In fact, the number of photons collected is proportional to the *area* of the mirror, rather than its diameter, so the improvement factor is bigger than you might think. Allowing for the hexagonal shape of the segments, and the big hole in the middle, the useable area of Webb's mirror is 25 square metres. That's more than six times that of Hubble, which has an effective area of around four square metres.

Apart from the difference in wavelength, Webb's optical design is very similar to Hubble's. It's essentially a Cassegrain telescope that uses a smaller secondary mirror to direct the focused image from the primary mirror through a hole in its centre to various detectors located behind it. In external appearance, however, Webb looks very different from Hubble, with a skeletal structure that – at first glance – resembles a radio telescope more than an optical one, although all that's missing is the cylindrical outer tube that encloses Hubble's optical assembly. Hubble needs this to shield its mirrors from stray light – which, from its vantage point in Earth orbit, it receives a lot of, from blazing sunshine in one direction to reflections from the planet's surface in another.

But Webb doesn't have to worry about any of that. Following in the footsteps of WMAP, Planck, Herschel and Gaia, it's located at that special point 1.5 million kilometres away known as L2 – the second Lagrange point. As you'll remember from the previous chapters, this has the advantage that the Sun and Earth are always in the same direction. So all Webb needs is a single sunshield, which – as we saw in the case of Gaia – can be oriented in such a way to give the spacecraft a permanently cold side and a permanently hot one. The latter carries solar panels for power and an antenna for two-way communication with Earth, while the cold side is the one that does the observing.

The sunshield separating the two sides is, in Webb's case, truly enormous. Shaped roughly like a kite and made up of five separate layers, it measures 21 metres long by fourteen metres wide – comparable in size to a tennis court. While the sunlit side may reach temperatures of 100° Celsius, the cold side hovers around –237° Celsius. That's a perfect operating temperature for an infrared telescope.

While L2 provides an ideal thermal environment for a space telescope, it does come with a downside. Its enormous distance from Earth – four times further away than the Moon – means it's not going to be easy to fly out to Webb and repair it if anything goes wrong. In fact, it was never designed to be serviced in space in the way that Hubble was. Instead, everything had to work perfectly first time. That would be a big ask even if we were talking about a small spacecraft that could be fitted neatly inside the nose cone of a launch rocket, but things were never going to be that simple with Webb. Remember that tennis-court-sized sun-shield? It had to be folded up to much smaller dimensions for launch, as did the 6.5-metre-diameter mirror. Unfolding all that technology in space, with no humans present – and expecting everything to 'work perfectly first time' – was an engineering nightmare of the first order.

Although I've been describing Webb as a NASA project, that's not strictly accurate. Like Hubble before it, it's actually a joint venture between NASA and ESA, with the latter very much the junior partner (and there's a small contribution from the Canadian Space Agency too). In return for a 15 per cent share of observing time, ESA supplied one of Webb's scientific instruments and was responsible for its launch on Christmas Day 2021.

Transforming Webb from its folded-up state to its final operating configuration was a slow and intricate procedure.

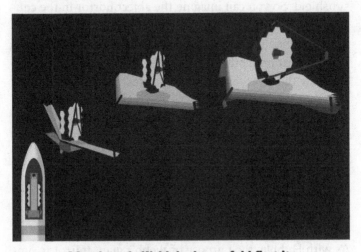

**After launch, Webb had to unfold first its sunshield and then its main mirror.**

ESA

It took up most of the time from the moment the spacecraft separated from the Ariane rocket, half an hour after launch, until it arrived at L2 almost a month later. The process involved '344 single points of failure', which means there were 344 mission-critical operations, every one of which had to happen exactly right. If just one of them had failed, the spacecraft would have been completely useless, with no prospect of a Hubble-style rescue mission. In other words, $10-billion-worth of space junk.

In the event, a small piece of hardware did fail during the deployment process. Fortunately, it wasn't one of the mission-critical ones, but – ironically – a sensor whose sole purpose was to tell ground control that one of the mission-critical steps had been completed correctly. This happened to be the very step that controllers had been most nervous about, the unfurling of Webb's intricately folded sunshield. So you can imagine the abject horror in the control room when they failed to receive confirmation that the unfolding had taken place successfully. It was only when it became clear from other data that the sunshield had deployed correctly that everyone could start breathing again.

Once past that brief glitch, the rest of the deployment sequence – and the methodical testing and calibration that followed – all went flawlessly. By the spring of 2022, Webb was ready to start operations.

## Webb's science mission

As with Hubble, the purpose of Webb's giant mirror is simply to collect photons and focus them down onto a number of different detectors – and it's the latter that do the real scientific work. In Webb's case, there are four separate science instruments:

- Near-Infrared Camera (NIRCam). Webb's primary imaging instrument, this operates, as its name suggests, in the near-infrared band. If you've seen any really spectacular images from Webb, it's a pretty safe bet they were taken with NIRCam. But it's designed to do serious science as well, with a wide remit ranging from the most distant galaxies to young stars in the Milky Way and small, faint objects within our own Solar System.

- Near-Infrared Spectrograph (NIRSpec). This provides spectroscopy over the same wavelength range as NIRCam. Capable of measuring the spectra of up to 200 objects at once, it will supplement NIRCam's imagery with information on temperature, mass and chemical composition. NIRSpec is ESA's main contribution to Webb's hardware.

- Mid-Infrared Instrument (MIRI). The only instrument covering the longer-wavelength end of Webb's range – in the mid-infrared band – this acts both as a camera and a spectrometer. MIRI requires a lower operating temperature than the near-infrared instruments, around six Kelvin (or –267° Celsius), which is provided by a special cooling system that uses helium gas as the refrigerant.

- Near-Infrared Imager and Slitless Spectrograph (NIRISS). This is the Canadian contribution to Webb, and it's coupled to another gadget – not part of the scientific suite – called the Fine Guidance Sensor, which is what allows Webb to point very accurately at a particular target. NIRISS itself complements NIRCam's near-infrared imagery and NIRSpec's spectroscopy in certain specialised areas – notably the analysis of exoplanet atmospheres.

Webb's science goals are organised into four themes, spanning some of the most important questions in contemporary astronomy. To start with, there's probably the biggest question of all: how did the universe begin? Like Hubble before it, Webb can be thought of as a 'time machine', thanks to the fact that it can see distant galaxies as they used to be long ago in the past. Hubble was able to peer back billions of years, but Webb will see even further – possibly even to the moment the first galaxies formed, around 13.6 billion years ago. One of its first tasks is to carry out a survey called COSMOS-Webb, which is effectively an even deeper version of Hubble's deep-field images.

The second of Webb's science goals is what NASA refers to as 'galaxies over time'. This again draws on Webb's role as a kind of time machine. We know that most nearby galaxies – which we see as they were relatively recently, and therefore in mature form – tend to be large and symmetrical. Astronomers have long assumed that this is the result of billions of years of evolution, and that far more distant – and therefore much younger – galaxies would be smaller and scrappier-looking. The tantalising glimpses of such galaxies that Hubble managed to obtain supported this view. But Webb is able to see these very young galaxies in even greater detail – and its first results came as a surprise, to say the least. The telescope found galaxies that look very like our own right out on the edge of the observable universe, where no one expected them to be. This caused turmoil in the theoretical community – which, of course, is exactly what a brand-new super-telescope ought to do.

The third of Webb's scientific aims deals with the lifecycle of stars. Unlike galaxies, which were all born very early in the history of the universe and have steadily got older since then, the stars inside them go through lifecycles – and a progressive sequence of generations – that's more akin to living creatures. They're born, develop, age and die, with the remnants of old stars contributing to the raw material that makes new stars. While astronomers already understand quite a lot about this process, there are still a few mysteries. That's particularly true when it comes to the actual birth of stars – and of any planets that may form around them – due to the enveloping cocoon of dust that surrounds these processes. It's all but impenetrable to visible-band telescopes like Hubble, but Webb should be able to see through it to reveal those ultimate secrets of star formation.

Finally, Webb's fourth science goal is the one which, speaking personally, I always find the most exciting – the one that NASA refers to as 'other worlds'. In fact, because I do find it so interesting, I've probably already given the game away in previous chapters by describing how Webb will use transmission spectroscopy to probe the chemical composition of exoplanet atmospheres in search of tell-tale 'biosignatures'. The telescope really hit the ground running on this one, with the first scientific paper detailing Webb's analysis of an exoplanet atmosphere – a collaborative effort by over 80 astronomers – appearing in February 2023. And Webb's exoplanetary studies go further, too. Its ability to see through clouds of dust, and its super-high imaging resolution, may also give us a view of planetary systems in the very process of formation.

# HIGH-ENERGY ASTRONOMY 7

The space telescopes we've encountered so far cover a large chunk of the electromagnetic spectrum. They span a wavelength range from the longest covered by WMAP – 1.3 centimetres in the microwave band – all the way down to Hubble's shortest wavelength – 100 nanometres in the ultraviolet. That encompasses pretty much all the thermally produced emissions that are likely to be of interest to astronomers, from the ultra-cold 2.7 Kelvin black-body radiation of the CMB to the sweltering 30,000° Celsius of the hottest blue stars. But this doesn't mean we've covered the whole range of phenomena that it's possible to explore with a telescope. There are other processes that produce even shorter wavelengths, in the X-ray and gamma ray bands.

As we saw in Chapter 1, shorter wavelengths correspond to higher photon energies. For much of the book, this has been a mere technicality, but as we progress up beyond the ultraviolet, the wavelengths start to get so tiny that our brains can't attach any meaning to them. Beyond a certain point, it makes more sense to talk in terms of the energies involved instead. By convention, the unit used here is the electron-volt (eV), which originated in the way X-rays are created in the machines found in hospitals. These use powerful electric fields to accelerate electrons up to high speed, which in turn causes the electrons to produce X-rays. The voltage needed to accelerate the electrons is typically a few thousand volts, and so the resulting photon energies can be expressed as a corresponding number of 'kilo-electron-volts' or keV.

While this set-up isn't generally used to produce lower-energy ultraviolet or higher-energy gamma radiation, the same measurement system can be applied to those bands too. So for example, a photon of ultraviolet radiation at 100 nanometres wavelength has an energy of 12.4 eV, while gamma rays can extend up into the mega-electron-volt (MeV) range, or even higher. On Earth, photons as energetic as this are only ever encountered in the context of subatomic processes, for example, in the radiation from nuclear weapons or radioactive materials.

Both types of high-energy radiation, X-rays and gamma rays, can be produced by astronomical processes – and so astronomers naturally want to study them with telescopes. You could be forgiven for thinking that because the photons involved have such high energies, they can tear unimpeded through the Earth's atmosphere and easily be detected by instruments on the ground. After all, the most familiar use of X-rays is to see through objects, isn't it? But as logical as this argument sounds, it couldn't be more wrong. The molecules in the atmosphere just love absorbing all that high-energy radiation, and almost none of it actually makes it to ground level. When you think how harmful an X-ray overdose can be, let alone exposure to gamma rays, that's obviously great news for us. But it does mean that anyone wanting to do astronomy at X-ray or gamma ray energies is going to have to do it from space.

## X-rays

The first attempts to detect X-rays from cosmic sources were made in the late 1950s, using suborbital rocket flights that could only spend a few minutes at a time above the atmosphere. This work was led by a young Italian–American astrophysicist named Riccardo Giacconi, and by 1962, his team had succeeded in detecting the first X-ray source outside the Solar System, Scorpius X-1.

This result was tantalising enough that Giacconi drew up plans for a purpose-built satellite to study such sources in more detail. He put the idea to NASA's chief of astronomy, Nancy Roman, and suggested that it might be included in the Explorer programme of small scientific satellites. This was a far from trivial request because back in those days – the early 1960s – building and launching even a modest-sized satellite was a significant undertaking. Fortunately, however, Roman quite liked Giacconi's proposal, as he later explained:

> It was a delightful surprise to hear Dr Roman express the opinion that NASA might, in fact, be interested in considering an X-ray Explorer fully devoted to a search for celestial X-ray sources.

By the standards of modern space telescopes, Giacconi's design really was very small, with a total mass of just 140 kilograms. In fact, strictly speaking, it wasn't a telescope at all because it didn't contain any optics to magnify and focus an image. Instead, it was simply an array of bare detectors with narrow fields of view – the idea being that, by rotating the satellite, it would be possible to work out the general position of a source in the sky. The detectors covered an energy range from 2 to 200 keV and were sensitive enough to detect sources 10,000 times weaker than Scorpius X-1.

Although Giacconi got his original proposal in to NASA as early as April 1963, it wasn't until 12 December 1970 that the satellite was actually launched into space. Unusually for NASA, the launch took place from Kenya, which was the best location from which to achieve the desired equatorial orbit. The launch site also had a consequence for the satellite's name – which was formally Explorer 42 but became much better known as Uhuru. This means 'freedom' in Swahili, the official language of Kenya, and 12 December just happens to be Kenya's Independence Day. The name also has pleasant echoes of Lieutenant Uhura, the popular character from *Star Trek* – whose name is derived from the same word.

In the course of just over two years of operation, Uhuru scanned the whole sky several times to produce a catalogue of 339 X-ray sources – far more than had been known previously. In conjunction with Giacconi's earlier work using suborbital rockets, this was enough to win him a Nobel Prize in 2002 'for pioneering contributions to astrophysics, which have led to the discovery of cosmic X-ray sources'. By that time, he was one of the grand old men of astronomy, having served as the director of NASA's Space Telescope Science Institute from its inception in 1981 to 1993, after the launch of the Hubble telescope that was its main raison d'être.

Many of the X-ray sources identified by Uhuru fall into a particular class of object – called X-ray binaries – in which the production of X-rays is so central to their nature that it's part of their name. These are binary star systems in which one member is a relatively normal star and the other is – well, about as far from normal as it's possible to get. It's so incredibly small and dense that it has a very powerful gravitational field at close range – and it orbits so close to the other star that it constantly pulls material off it. As this material falls onto the small, dense object, it's heated up to the enormously high temperatures needed to produce X-rays.

As for the small, dense object itself, there are two possibilities for what it can be. The first is a neutron star – which is the case with Scorpius X-1, for example. Neutrons are one of the three types of subatomic particle, along with protons and electrons, that make up the atoms of ordinary matter. The protons and neutrons are found inside the tiny nucleus of the atom, while the electrons orbit much further out – accounting for most of the physical size of the atom. A neutron star is basically a star that's collapsed down under its own gravity to the point where all the electrons and protons have been squished together into further neutrons, so you're left with nothing but neutrons. If this happened to our own Sun,* the resulting neutron star would be about twenty kilometres in diameter.

---

* Which it won't because it isn't massive enough.

Of the two possibilities I mentioned, a neutron star is the less extreme. As for the other possibility – I'm willing to bet you already know what it is because it's the one type of exotic astronomical object that pretty much everyone's heard of. It's a black hole, which is what happens when a star continues to collapse past the neutron star stage all the way down to a 'singularity' – an infinitesimally small point containing the entirety of the star's mass. It's a concept that was theorised about for years before anyone had any idea if they really existed or not. The first hint that they did came from Uhuru.

One of the sources that Uhuru studied was called Cygnus X-1. Almost immediately, this stood out as something of an oddity because the strength of the X-rays it was emitting fluctuated very rapidly – several times per second, in fact. To astrophysicists like Giacconi, this could only mean that the X-rays were emanating from a very small region – yet one that seemed to enclose a very large mass. Did this mean Cygnus X-1 contained a black hole? Follow-up observations by other instruments strongly suggested this was the case, although other explanations were still possible.

The ambiguity over the nature of Cygnus X-1 had an amusing consequence. One of the many scientists who had put a lot of time and effort into the theoretical study of black holes was Stephen Hawking, and unsurprisingly, he really wanted Cygnus X-1 to be a black hole. As 'a form of insurance policy', as he put it, he made a bet in 1974 with his colleague, Kip Thorne, that Cygnus X-1 *wasn't* a black hole. If this was subsequently proven to be the case, then at least Hawking would have the consolation of winning his side of the bet – a four-year subscription to the British satirical magazine *Private Eye*. But that never happened, and Hawking eventually conceded the bet and paid up.

In 1976, a few years after the end of the Uhuru mission, Giacconi came up with another proposal, this time for a much larger and more ambitious instrument that was initially dubbed the Advanced X-ray Astrophysics Facility, or AXAF. This soon found its way into NASA's 'great observatories' programme – but like the Hubble telescope that featured in the same programme, it had to endure a long process of planning, development and schedule delays before it actually made it into orbit.

The X-ray component of the 'great observatories' programme was finally blasted into space in July 1999, in the payload bay of the Space Shuttle Columbia. With a total mass of more than 22 tonnes, it was the largest satellite ever launched by a Shuttle – and the mission made the history books for another reason, too. The left seat on the flight deck was occupied by NASA astronaut Eileen Collins – the first time a Shuttle mission had a female commander.

By this time, the satellite been renamed Chandra, after the Indian astrophysicist Subrahmanyan Chandrasekhar. Although he worked on many different topics, he's perhaps best known for calculating what is now referred to as the 'Chandrasekhar limit' – the maximum mass that a star can have before it undergoes uncontrolled gravitational collapse at the end of its life. When he worked the theory out in the 1930s, Chandrasekhar didn't know what the end result of this collapse would be; he only knew that it had to happen. Today, we know the collapse leads to one of the two states found in an X-ray binary, either a neutron star or a black hole. So giving Chandrasekhar's name to an X-ray telescope is entirely appropriate. For his 'theoretical studies of the physical processes of importance to the structure and evolution of the stars', Chandrasekhar was awarded a 50 per cent share in the 1983 Nobel Prize in physics.

Besides Hubble, the only other member of the 'great observatory' programme we've met so far is the Spitzer infrared telescope, and as I said in the previous chapter, it's questionable whether that really deserved the adjective great. In terms of its size and technical specification, Spitzer was really only an average specimen as space telescopes go. But I can't make the same comment about Chandra, which is a true counterpart of Hubble in every way. At 13.8 metres in length, it's actually slightly larger than Hubble, and it has produced results that are just as spectacular in its own field. Along with Hubble, too, it's the only one of the four observatories that's still operational as I write this in mid-2023. Sorry if that was a bit of a spoiler – because we haven't met the fourth member of the great telescopes yet – but I'm only anticipating events by a few pages, as it's going to make an appearance before the end of this chapter.

A little like TESS, the planet-hunting satellite we met in Chapter 4, Chandra was placed in a highly elongated orbit around the Earth – although not quite such an enormous one in this case. At its closest point, Chandra's orbit is about 16,000 kilometres above the Earth's surface, and at its furthest about 133,000 kilometres away – around a third of the distance to the Moon. The reason Chandra needs such a big orbit is to keep it clear of the intense radiation belts that surround the Earth, which would overwhelm the telescope's sensitive detectors. As it is, Chandra takes 64 hours to complete a single orbit, and during 55 of those hours it's high enough to be clear of the radiation belts and able to make uninterrupted observations. For the nine hours that it swings through its lowest altitudes, its instruments are switched off to protect them.

Unlike Uhuru, Chandra is a 'real' telescope, which uses mirrors to focus the incoming X-rays to form a sharp image. But the laws of optics work slightly differently at X-ray wavelengths as opposed to the visible, infrared and ultraviolet bands, which means the optimum geometry for the mirrors is different too. Instead of arranging things so the incoming rays hit the mirror more or less perpendicularly, the best way to get a good X-ray reflection is to have the rays hit the mirror at a much shallower angle. So as odd as it looks, Chandra actually uses a series of nested mirrors shaped like tapered cylinders, which are placed at the front end of the tube – rather like the lens in a refractor telescope – instead of the back, as in a traditional reflector design.

With an effective aperture of over a metre, Chandra has eight times better resolution and up to 50 times greater sensitivity than any previous X-ray observatory. By staring at the

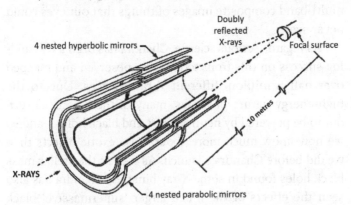

**NASA's Chandra space telescope uses an unusual arrangement of mirrors optimised for the X-ray band.**

NASA

same spot for a very long time – albeit with those enforced nine-hour breaks – it can pick up incredibly faint sources, the weakest of them equivalent to receiving just one photon every four days. In terms of photon energy, Chandra has a range from around 0.2 to 10 keV – or between about 0.1 and 6 nanometres expressed as a wavelength. The first of those figures is about the size of an atom, while the larger one is still somewhat smaller than a DNA molecule.

These specs give Chandra an unprecedented ability to produce crystal-clear images at X-ray wavelengths. For this reason, some of its results have filtered into the public arena a little more noticeably than, say, Uhuru's did. The high-resolution photographs it's taken of supernova remnants, colliding galaxies and the like – once they've been suitably colourised by NASA's PR people – can rival anything Hubble has produced. In fact, like Spitzer in the infrared band, Chandra's X-ray images have sometimes been superimposed onto Hubble's to produce fascinating, multi-band composite images of things that our eyes could never see.

As regards serious science, Chandra has been a resounding success on this front too, having observed and mapped over half a million different X-ray sources. Due to the high-energy nature of X-rays, many of these sources turn out to be powered by neutron stars and black holes – and so we now know much more about these exotic objects than we did before Chandra's launch. As well as the stellar-mass black holes found in some X-ray binaries, Chandra has also seen the effects of the much larger 'supermassive' black holes that power the nuclei of active galaxies and quasars. Although it's not as hugely active as some, there's even one of these supermassive objects at the heart of our own Milky

Way galaxy. Known rather cryptically as Sagittarius A* (that asterisk is part of its name, not a footnote reference), this has a mass 4.5 million times greater than the Sun, all crammed into a space smaller than the orbit of Mercury. While looking at Sagittarius A* in October 2013, Chandra observed the largest X-ray flare ever seen from that direction, around 200 times brighter than usual. One theory is that the black hole had just gobbled up a small asteroid-sized object that strayed too close to it.

### Gamma rays

If an asteroid falling into a black hole can cause such a massive outburst of X-rays, what's going to happen when a whole star – sufficiently far above the Chandrasekhar limit – collapses down to a black hole? It's a process that may only take a matter of seconds, yet it releases a tremendous amount of gravitational energy. The result is a sudden burst of gamma rays, at even higher photon energies than X-rays – all the way into the MeV (mega-electron-volt) range.

These awesome phenomena, known as gamma ray bursts (GRBs), were discovered by satellites – but not astronomical ones. They were operated by the US Air Force, which launched a series of small satellites called Vela in the 1960s. These were designed to look for sudden bursts of gamma rays with a much more down-to-earth origin, in the form of the nuclear weapon tests that had been banned by the Nuclear Test Ban Treaty in 1963. The Vela satellites didn't uncover any treaty violations, but they did detect occasional flashes of gamma rays coming from an entirely unexpected direction: outer space.

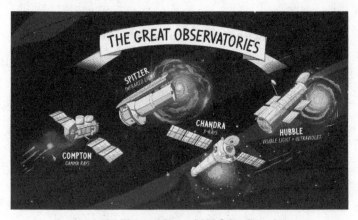

**A cartoony illustration of NASA's four 'great observatories': Hubble, Spitzer, Chandra and Compton.**
NASA

For a long time, the origin and nature of GRBs remained a mystery to astronomers, and one that various small astronomical satellites did little to clear up in the 1970s and 1980s. The serious investigation of GRBs had to wait for the high-energy member of NASA's 'great observatories' programme, the Compton Gamma Ray Observatory.

The new telescope was named after the American physicist Arthur Compton, who worked on gamma rays among other topics. His spacefaring namesake was the first large telescope designed for gamma ray astronomy – and it really was large, its seventeen-tonne launch mass easily putting it in the same class as Hubble and Chandra.* Compton was launched into low Earth orbit by the Space Shuttle Atlantis in April 1991, twelve months after Hubble. It operated successfully for nine years, until a problem with its gyroscopes

---

* I won't knock Spitzer again, other than to remind you that its launch mass was a mere 950 kilograms.

put it at risk of an uncontrolled crash back to Earth. In the interests of safety, NASA deliberately 'deorbited' Compton into the Pacific Ocean on 4 June 2000.

Gamma ray energies are so high that even the low-angle reflections used in X-ray telescopes won't work. This means that telescopes like Compton can't employ an ordinary optical system to focus incoming gamma rays, so they have to find another way to produce a sharp image. A range of technologies have been developed to do this, such as 'collimation'. If that word doesn't mean anything to you, imagine looking through a straight, narrow tube. It will limit your field of view to a very localised area – and that, basically, is how collimation works. I ought to add that the collimators used in gamma ray telescopes are rather more sophisticated than this – but you probably guessed that already.

Compton carried a suite of four scientific instruments, containing both imagers and spectrometers, with more than ten times the sensitivity of any previous gamma ray satellite. Between them, the instruments spanned a huge range of photon energies, all the way from twenty keV up to 30 GeV, where GeV stands for 'giga-electron-volt'. That means a billion electron-volts, or 1,000 MeV – which, when you're talking about processes on the minuscule scale of an atomic nucleus, really is a lot of energy. If you took a proton and converted its mass into pure energy using Einstein's $E = mc^2$ formula, you'd only end up with 0.94 GeV.

As well as short-lived bursts, Compton was designed to map the locations of more persistent gamma ray sources – often the same super-hot astronomical objects that Chandra had mapped in the X-ray band. Compton was responsible for the first survey of the whole sky at gamma ray energies, in the course of which it discovered a whole new class of active galaxies that emit most of their radiation at these ultra-high energies.

As for GRBs themselves, Compton's main instrument devoted to their study was the 'Burst and Transient Source Experiment', or BATSE. This made two critical discoveries that completely changed the way astronomers thought about GRBs. For one thing, it became clear that they're a much more common occurrence than previously thought. In the course of nine years of operation, BATSE detected an average of one burst every day.

BATSE's second key discovery was that GRBs are equally likely to come from anywhere in the sky, rather than being confined to the plane of the Milky Way. Up to this point, astronomers had imagined that – because the GRBs we see are so powerful – they must be coming from relatively nearby objects, such as neutron stars, inside our own galaxy. But that clearly wasn't the case – the GRBs that BATSE detected had to originate much further away in the universe.

After careful analysis of Compton's results, as well as those from later follow-on missions, astronomers were able to build up a much better understanding of the GRB phenomenon. In fact, it turns out there are two distinct types of burst. The first type, which generally last less than a couple of seconds, are most likely to be caused when pairs of black holes or neutron stars collide and merge together. This isn't something that has a high probability of occurring, but you have to remember that the universe is so vast that even low-probability events are happening somewhere all the time. And when they do, they produce so much energy that there's a chance we'll see it in the form of a GRB. In fact, the same massive-object mergers also release energy in another form, as gravitational waves – which, as we'll see in the next chapter, gives us another way to detect them.

The second type of GRB lasts longer, with a duration ranging from two seconds to several minutes. These really do come from the situation I described at the start of this section when a massive star comes to the end of its life and collapses down to a black hole. Under some circumstances, this collapse can result in what's known as a 'hypernova' explosion, and that's what produces the sudden flash of high-energy gamma rays. As the name suggests, a hypernova is an extreme form of supernova, a hundred times more powerful than the ordinary kind (which don't produce GRBs). As with black hole mergers, hypernovas aren't a common occurrence, but the gamma ray flashes they produce are so intense that we can detect them even if they happen billions of light years away, close to the edge of the observable universe.

If you think about it for a second, there's a rather alarming corollary to this. We've never actually seen a GRB that originated inside our own galaxy, but what if one did? A hypernova-type GRB can throw out as much energy as our own Sun will produce in the entirety of its 10-billion-year lifetime – all in a matter of seconds. If something like that happened within a few thousand light years of the Solar System, the sudden blast of radiation would be devastating for life on Earth. It would almost certainly result in the extinction of our species. I'd like to give you a nice comforting reason why a nearby GRB will never happen, but the fact is that it might. The best we can say is that – as with a devastating asteroid impact – it's statistically unlikely to happen within the lifetime of the human species. All the same, you might like to add 'gamma ray burst' to your list of possible end-of-the-world scenarios.

As with all the other wavebands we've covered in this book, the technology available for gamma ray telescopes is improving all the time. Currently, the one with the highest spec is NASA's Fermi space telescope – named after another pioneer of high-energy physics, Enrico Fermi – which was launched in 2008 with the primary aim of searching for GRBs. Not long ago, I tried to impress you by saying that Fermi's predecessor, Compton, could detect photons up to energies of 30 GeV. But Fermi can go ten times better than that, all the way up to 300 GeV. That's comparable to the kinetic energy of a mosquito – which is definitely enough to annoy you, if it hit you in the face – except that it belongs to a single photon of radiation, rather than a centimetre-sized insect.

Fermi is also capable of pinpointing the location of GRBs more accurately than its predecessor, in both the space and time dimensions. It can measure the arrival time of a burst to within a few microseconds and detect very brief bursts even if their total duration is only a matter of microseconds. As for its angular accuracy – well, after so many superlatives, you may be a little disappointed here. The lack of optical focusing in gamma ray telescopes means they're never going to have the sub-arcsecond resolution of Hubble, Webb or Gaia. In fact, by the standards of its field, Fermi does extremely well with an accuracy of 60 arcseconds, or about a thirtieth of the diameter of a full Moon.

If Fermi represents the current state-of-the-art, what comes next? One possibility is an ESA proposal called THESEUS, which is worth mentioning if only because I think it's the cleverest of ESA's many contrived mission names

I've ever come across. It stands for 'Transient High-Energy Sources and Early Universe Surveyor', which is a pretty accurate description of what it's designed to do. The telescope's basic aim is to detect GRBs and other transient high-energy phenomena all the way out to the very edge of the visible universe. Using a unique suite of instruments, it would not only detect short-lived X-ray and gamma ray emissions, but immediately follow them up with infrared observations to link the detection to a particular host galaxy.

As exciting as that sounds, at the moment, THESEUS is nothing more than a proposal on paper. If it ever gets off the ground, it won't be until the mid-2030s at the earliest. This puts it firmly in the future – which just happens to be the subject of our next and final chapter.

# THE FUTURE    8

In the course of this book, we've encountered space telescopes spanning virtually the whole of the electromagnetic spectrum. Running through the list in order of descending wavelength, there were three generations designed to study the cosmic microwave background, COBE, WMAP and Planck, followed by the infrared telescopes Spitzer, Herschel and Webb. Specialist instruments in the visible waveband included the planet-hunters Kepler and TESS, and galaxy-mappers Hipparcos and Gaia. The most famous space telescope of all, Hubble, spans a range of wavelengths including the short end of the infrared, the whole of the visible and most of the ultraviolet. At even shorter wavelengths – and consequently higher photon energies – came Uhuru and Chandra in the X-ray band, and the gamma ray telescopes Compton and Fermi.

Between them, these instruments have made tremendous contributions to the science of astronomy, covering the whole field from the diversity of planets and the lifecycles of stars to the structure of galaxies and the evolution of the universe. Without them, we would know far less than we do about exoplanets, black holes and the cosmic microwave background – and perhaps nothing at all about topics like dark energy and gamma ray bursts.

Ever since telescopes were first invented, every step forward in technology – whether that's a bigger aperture, a different waveband or an improved type of detector – has led to new discoveries. Sometimes these discoveries have solved longstanding mysteries, but as often as not, they've raised whole new mysteries of their own. The upshot is that astronomy can never stand still, and there's always a demand for bigger and better telescopes.

So what does the future hold? I've already mentioned a few of the new space telescopes that are being planned, such as the PLATO and ARIEL exoplanet hunters, which will expand on the work already done by Kepler and TESS, and the THESEUS proposal for an advanced high-energy telescope to probe gamma ray bursts. These three projects all come from ESA, and there's another one from the same agency that's worth a brief mention too. It's called Euclid – and, in ESA's words, it's 'designed to explore the evolution of the dark universe'.

Euclid isn't an abbreviation for anything – it's the name of an ancient Greek mathematician who was a pioneer in the field of geometry (ESA's rationale seems to be that the nature of dark matter and dark energy are closely linked to the geometry of the universe). Euclid's aim is to map the distribution of billions of galaxies out to a distance of 10 billion light years, to give a more precise picture of how the structure of the universe has evolved over time. The hope is that this improved dataset will allow a better understanding of the nature of dark matter and the origin of dark energy – if, indeed, the latter really does exist, and isn't just an illusion caused by measurement errors.

As contemporary space telescopes go, Euclid is relatively modest in size, with a mass of two tonnes and an aperture of 1.2 metres. But it will have two state-of-the-art instruments, an imaging camera in the visible band and a combined camera and spectrometer covering near-infrared wavelengths. Launched on 1 July 2023, Euclid will conduct its scientific mission – anticipated to last at least six years – from the ever-popular L2 point.

On the other side of the Atlantic, NASA's next big space telescope is also in the advanced stages of development. Originally called WFIRST, which stood for 'Wide-Field Infrared Survey Telescope', it was renamed following the death of NASA's first chief of astronomy – and 'mother of Hubble' – at the grand old age of 93 in 2018. So it's now the Nancy Grace Roman Space Telescope, or just 'Roman' for short. In some ways, Roman can be considered a more direct replacement for Hubble than Webb, as it covers a similar range of wavelengths in the near-infrared and visible parts of the spectrum, and with a main mirror that's made of glass and 2.4 metres in diameter.

At this point, it's worth recalling a kind of 'urban myth' about Hubble that you may have heard. It comes in various forms, but they all seek to relate Hubble's design to the spy satellites used by the United States during the Cold War – for example, that Hubble's 'faulty' mirror was a reject from one such satellite. As far as Hubble is concerned, the only grain of truth in these legends is that the spy satellites did indeed use the same diameter of mirror, 2.4 metres, but they were designed to look at locations that were only a few hundred kilometres away, on the surface of the Earth, rather than distant astronomical objects. So the optical design of the mirrors was completely different; in particular, the spy satellites had a much shorter focal length than Hubble.

Why am I telling you this now? Because the urban legend essentially comes true with Roman. The telescope really did start life as a spy satellite, manufactured for the US National Reconnaissance Office, but it was donated to NASA when it became surplus to requirements (although hopefully not because of a faulty mirror). Its short focal length means that, when applied to astronomy, the telescope will have an unusually wide field of view, as reflected in the first two letters of its original name, WFIRST. You may remember that one of Hubble's scientific instruments is also described as a 'wide-field' camera – but that's something of a misnomer, as its angular field of view is only 164 arcseconds across, or less than a twentieth of a degree. In contrast, Roman's field of view really is wide – by astronomical standards – at 0.53 degrees, slightly larger than the full Moon.

Aside from its wider field of view, Roman's optical spec will be very close to Hubble's. With the same aperture size, it will have the same sensitivity and resolution as its illustrious predecessor, but it will be able to image a greater number of objects simultaneously. This makes it ideally suited to carrying out wide-area surveys in a comparatively short time, and its primary science focus will be on two fashionable areas of astronomy where such surveys are particularly useful. At the risk of sounding like a broken record, these are exoplanets – the same as ESA's PLATO and ARIEL – and the 'cosmological' topics of dark matter and dark energy, like ESA's Euclid. Unlike those specialist telescopes, however, Roman has a broader remit as well, comparable to Hubble and Webb. Like the latter, it will be located at the L2 point, and it's likely to be launched towards the end of the 2020s.

As we've seen throughout the book, there's more than one reason for putting a telescope in space. In the case of microwave instruments like COBE, WMAP and Planck, the main motivation was to get away from terrestrial emissions at the same wavelength that would swamp the tiny signals they were looking for. With infrared telescopes, it's because the Earth's atmosphere blurs and scatters the incoming waves to the point that it's difficult to form useful images of astronomical objects. And in the case of X-rays and gamma rays, they hardly penetrate the atmosphere at all.

With visible light, the reasons aren't so obvious. For one thing, we know that such light happily passes through the atmosphere, because we can see the stars – and even an amateur astronomer can take bright, clear photographs of them. But skyglow limits the faintest objects that can be seen, while the constant movement of air molecules in the atmosphere causes stars to wobble slightly – or twinkle – which makes it difficult to obtain really sharp pictures of them. There's nothing we can do about the first of these problems – so we're always going to need space telescopes like Hubble for really 'deep field' images at visible wavelengths – but what about the second one?

The sledgehammer way to produce sharper images of stars is simply to build a bigger telescope, and that's something that's always going to be simpler to do at ground level than in space. And with improvements in technology, it's getting easier all the time. When Hubble was designed in the mid-1970s, the largest telescope in the world was the '200-inch' reflector on Mount Palomar in California, which has an aperture of 5.1 metres – only a little over twice that of Hubble. And like Hubble's, the Palomar mirror was made from a single piece of glass. Today, on the other hand, there are a couple of ground-based telescopes under construction that are going to make Palomar look like something an ambitious amateur astronomer might try to impress their neighbours with.

This is a book about space telescopes, so perhaps I shouldn't speak up too much for the competition – but I'm talking here about ground-level observatories that should produce images with ten to sixteen times better resolution than Hubble. One is called the Giant Magellan Telescope, with an aperture of 24 metres, and the other is – appropriately enough – the 'Extremely Large Telescope', with an even greater aperture of 39 metres. They both use the same combination of tricks to achieve their Hubble-beating performance. The first is down to their location, which in both cases is in Chile. The combination of high altitude and very dry conditions means that the Chilean night skies are as sharp and clear as anywhere on Earth. The second trick is to use, instead of a single huge mirror – which would be totally impractical for the dimensions involved – an array of smaller elements, similar to Webb. This approach was forced on Webb because its mirror had to be folded up for launch, but down here on terra firma it has another advantage. There's really no limit, apart from the cost, to the size of telescope you can build this way.

The third trick the new telescopes use is pure technology, and something that would have seemed like magic not very long ago. The secondary mirrors – the ones that collect light from the main mirrors and direct it down onto the scientific instruments – are flexible, with their precise shapes at any moment controlled by hundreds of electronically operated actuators. These constantly tweak the secondary mirrors in just the right way to counteract the effects of atmospheric turbulence, thus transforming a twinkling star into a sharp point of light as steady as anything Hubble sees. While it would be an exaggeration to say that this technique – known as adaptive optics – spells the death of space telescopes at visible wavelengths, it does mean that high-end ground-based telescopes are going to be increasingly competitive in terms of the information they can collect compared to space telescopes.

Still, there are plenty of other wavebands where the desired performance is only ever going to be achieved by a space-based instrument. And one of them takes us outside the electromagnetic spectrum altogether.

## Gravitational waves

The idea of 'gravitational waves' made a cameo appearance in the previous chapter, as an alternative way to detect the mergers of black holes or neutron stars that produce short-duration gamma ray bursts. Like gravitational lensing and the Big Bang model, gravitational waves are one of the consequences of Einstein's theory of General Relativity – and as with most ideas connected with that theory, they're not the simplest thing to explain. For a proper description of what they are and how they can be detected, the best I can do is refer you to Brian Clegg's book about them in the Hot Science series.* All I'll do now is summarise the bare bones of the subject – which, hopefully, is all we're going to need.

---

* Brian Clegg, *Gravitational Waves: How Einstein's Spacetime Ripples Reveal the Secrets of the Universe* (London: Icon Books, 2018).

In the everyday world, we think of gravity as a force of attraction towards a massive object – in our case, the Earth. On the other hand, General Relativity sees gravity as a geometric distortion of 'space-time' (the four-dimensional continuum made up of the three dimensions of space and one of time). The more massive an object is, the more it distorts the space-time around it. Under certain circumstances, the interaction between very massive objects – such as those black hole and neutron star mergers – can create wave-like ripples that spread out through space-time in a kind of gravitational analogue of electromagnetic waves. And like their electromagnetic counterpart, these gravitational waves can carry a lot of information that's useful to astronomers.

The first detection of gravitational waves was announced in February 2016, around a century after the theory that predicted their existence. This long gap reflects just how difficult they are to pick up, but as soon as the technique had been mastered, more and more gravitational wave detections followed. Most of them originated in the same kind of massive object mergers that produce short-duration gamma ray bursts – which, as I said in the previous chapter, are far from being common events. But gravitational waves, like electromagnetic ones, can travel over unlimited distances, so we can potentially detect them even if they come from very distant parts of the universe.

That first discovery of gravitational waves was made with a pair of huge ground-based facilities – or 'telescopes', if you want to stretch the definition of the term – called the Laser Interferometer Gravitational-Wave Observatory (LIGO), located at two separate sites in the United States. The most obscure word in that name, 'interferometer', is one that we encountered on a previous occasion as the original meaning of the 'i' in the name of ESA's Gaia space telescope. But I didn't bother to explain what it meant, because in the end, Gaia wasn't an interferometer. And I'm not going to explain it properly now, either, because it's only indirectly relevant to the detection of gravitational waves (for fuller details, see Brian Clegg's book). Basically, an interferometer is an optical tool, which in LIGO's case makes use of laser beams. Its usefulness comes from the fact that a laser interferometer can measure extremely small displacements, such as those caused by a passing gravitational wave.

The LIGO installations are massive feats of engineering, with the laser beams travelling along two vacuum-filled tubes, each four kilometres long, arranged in a giant L shape. But as big as they are, they're still nowhere near big enough to detect all the gravitational wave signals of interest to astronomers. To see why this is, we have to consider the full spectrum of possible gravitational waves. As with its electromagnetic counterpart, we could characterise this in terms of wavelength or frequency, but those aren't the most convenient measures to use in this case. LIGO's detections correspond to wavelengths of thousands of kilometres, or frequencies of a few hertz, but other astronomical phenomena – beyond LIGO's capability – can push those figures to billions of kilometres or tiny fractions of a hertz. Measurements like that aren't terribly meaningful to most people.

Instead of counting the number of wave crests arriving per second, to get a frequency in hertz, it's more useful to think in terms of the *time* – in seconds, minutes or hours – between the arrival of wave crests. This is referred to as the 'wave period', and it gives us a more natural way of characterising gravitational waves than trying to get our heads round their enormous wavelengths or tiny frequencies. The wave period also has the benefit of being related more directly to the physical phenomena producing the waves, such as the orbital period of a compact binary system. For the black hole and neutron star pairs detected by LIGO, this can be as short as a few milliseconds, but for larger objects – such as the supermassive black holes that power the active nuclei of galaxies – it's going to be much longer, possibly several hours.

There are two reasons why ground-based systems like LIGO are never going to detect gravitational waves on this kind of timescale. First, there are the natural vibrations of the Earth itself, caused by seismic activity, that would swamp any faint extraterrestrial signals. Second, the sensitivity of a laser interferometer to long-period vibrations improves with the length of the arms, which really need to be millions of kilometres long rather than the few kilometres of LIGO. So, if

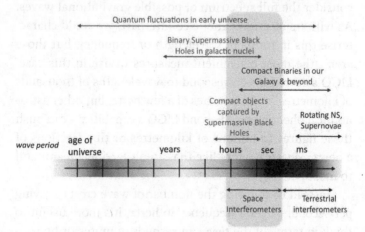

**The spectrum of gravitational waves as a function of wave period, showing the regions accessible to ground-based and space-based observations.**

NASA

you want to detect gravitational waves from supermassive objects, there's only one place to do it – and that's in space.

There have been several proposals for space-based gravitational wave detectors, but the one that's most likely to make it off the ground first is ESA's Laser Interferometer Space Antenna, or LISA. This device takes advantage of another huge benefit of outer space – the fact that it's already a vacuum. This means the interferometer arms don't need to be enclosed inside physical tubes – they can just be bare laser beams. As a result, LISA can be a truly gigantic affair, while requiring comparatively little actual hardware. ESA's proposal envisions three small spacecraft arranged at the corners of an equilateral triangle, each side of which is 2.5 million kilometres long. But the sides themselves simply consist of laser beams linking the three spacecraft together.

If you cast your mind back to previous occasions when we've touched on orbit theory, you'll realise that those three spacecraft can't just be put anywhere or they'd drift apart and the neat equilateral triangle would soon get broken. Kepler's laws of motion mean they have to orbit the Sun at exactly the same distance, to ensure they go round it at the same speed. And they can't be located too close to the Earth or their orbits would be affected by its gravity as well. The solution that ESA came up with involves the three spacecraft travelling around the Sun at the same distance as each other, but with different orbital inclinations, arranged in such a way as to constantly maintain the equilateral shape. Their distance from the Sun is the same as the Earth's, so they take exactly a year to complete an orbit – but they'll be well clear of the Earth itself, trailing some 50 million kilometres behind it.

While the LISA concept looks great on paper, it depends on several technologies that are brand new in the context of space science. For this reason, ESA launched a small-scale test vehicle, called LISA Pathfinder, as a proof-of-concept demonstrator in 2015. This successfully verified all the techniques it was designed to, giving ESA the confidence to go ahead with the full-blown LISA mission – although it still probably won't happen until the late 2030s.

When it does go ahead, LISA should be able to detect gravitational waves with periods up to about 1,000 seconds, or just over a quarter of an hour. This would allow it to observe processes involving supermassive black holes, and – even more excitingly – maybe even the long-period 'primordial gravitational waves' that cosmologists theorise were generated in the first moments after the Big Bang. This would push our knowledge of the universe's origin even further back than the cosmic microwave background, which represents the current limit to our observations.

But all that lies more than a decade in the future. Long before then, we can expect a whole range of amazing images and intriguing discoveries, in fields from exoplanets to cosmology, thanks to Webb, Roman, PLATO, Euclid and a host of other current and soon-to-be-launched space telescopes.

# FURTHER READING

## Chapter 1: Space and Telescopes

Francis Graham-Smith, *Eyes on the Sky: A Spectrum of Telescopes* (Oxford: Oxford University Press, 2016)

Andrew May, *The Telescopic Tourist's Guide to the Moon* (Cham, Switzerland: Springer, 2017)

'Lyman Spitzer: Making Space For Hubble', NASA, 30 July 2021 https://www.nasa.gov/feature/goddard/2021/lyman-spitzer-making-space-for-hubble

'Oral History Interviews: Nancy G. Roman', American Institute of Physics, https://www.aip.org/history-programs/niels-bohr-library/oral-histories/4846

'Spectroscopy: Reading the Rainbow', Space Telescope Science Institute, https://hubblesite.org/contents/articles/spectroscopy-reading-the-rainbow

## Chapter 2: Hubble

Jim Bell, *Hubble Legacy: 30 Years of Discoveries and Images* (New York: Sterling, 2020)

Brian Clegg, *Dark Matter and Dark Energy: The Hidden 95% of the Universe* (London: Icon Books, 2019)

R. Cowen, 'The Art and Science of Hubble's Images', *Eos*, 27 April 2015, https://eos.org/features/the-art-and-science-of-hubbles-images

Andrew May, 'Celebrating 30 years of the Hubble Space Telescope', *All About Space*, 25 April 2020, https://www.space.com/hubble-space-telescope-turns-30.html

'Gravitational Lensing', Stanford University, https://kipac.stanford.edu/research/topics/gravitational-lensing

## Chapter 3: Probing the Big Bang

Rhodri Evans, *The Cosmic Microwave Background: How It Changed Our Understanding of the Universe* (London: Springer, 2015)

Andrew May, 'The Big Bang Theory', *All About Space*, issue 126, February 2022

NASA Case Study: 'Redesigning the Cosmic Background Explorer', https://spacese.spacegrant.org/uploads/Requirements%20Config/COBE_case_study.pdf

'Planck and the Cosmic Microwave Background', ESA, https://www.esa.int/Science_Exploration/Space_Science/Planck/Planck_and_the_cosmic_microwave_background

'What is a Lagrange Point?', NASA, 27 March 2018, https://solarsystem.nasa.gov/resources/754/what-is-a-lagrange-point/

## Chapter 4: Exoplanet Hunters

Lucas Ellerbroek, *Planet Hunters: The Search for Extraterrestrial Life* (London: Reaktion Books, 2017)

Andrew May, *Astrobiology: The Search for Life Elsewhere in the Universe* (London: Icon Books, 2019)

Nadia Drake, 'Mystery of Alien Megastructure Star Has Been
Cracked', *National Geographic*, 3 January 2018, https://www.
nationalgeographic.com/science/article/mystery-of--alien-
megastructure--star-has-been-cracked

Michael Richmond, 'Transit-seeking satellites: Kepler and TESS',
http://spiff.rit.edu/classes/phys106/lectures/tess/tess.html

Govert Schilling, 'Transiting Exoplanet Survey Satellite: the
Mission to Find New Worlds', *Sky at Night*, https://www.
skyatnightmagazine.com/space-missions/tess-to-impress/

'The Habitable Exoplanets Catalog', University of Puerto Rico,
https://phl.upr.edu/projects/habitable-exoplanets-catalog

## Chapter 5: Mapping the Galaxy

Andrew May, 'How to Map the Milky Way', *How It Works*,
issue 130, October 2019

Andrew May, 'What is a Parsec?', Space.com, 29 July 2022,
https://www.space.com/parsec

'Gaia Data Release 3', ESA, https://www.cosmos.esa.int/web/
gaia/dr3

'Hipparcos Overview', ESA, https://www.esa.int/
Science_Exploration/Space_Science/Hipparcos_overview

'Hubble and Gaia Team Up to Fuel Cosmic Conundrum', NASA,
12 July 2018, https://www.nasa.gov/feature/goddard/2018/
hubble-and-gaia-team-up-to-fuel-cosmic-conundrum

## Chapter 6: Webb

Lewis Dartnell, 'James Webb Space Telescope is revealing
active exoplanet atmospheres', *BBC Sky at Night* magazine,
9 February 2023, https://www.skyatnightmagazine.
com/space-science/james-webb-space-telescope-
active-exoplanet-atmospheres/

Donna Lu, 'James Webb image reignites calls to rename telescope amid links to LGBT abuses', *Guardian*, 12 July 2022, https://www.theguardian.com/science/2022/jul/12/james-webb-image-reignites-calls-to-rename-telescope-amid-links-to-lgbt-abuses

Andrew May, 'James Webb Space Telescope: Origins, design and mission objectives', Live Science, 29 July 2022, https://www.livescience.com/james-webb-space-telescope

'History of Infrared Astronomy', ESA, https://www.esa.int/Science_Exploration/Space_Science/Herschel/History_of_infrared_astronomy

Tereza Pultarova, 'The James Webb Space Telescope discovers enormous distant galaxies that should not exist', Space.com, 22 February 2023, https://www.space.com/james-webb-space-telescope-giant-distant-galaxies-surprise

'Webb vs Hubble Telescope', NASA, https://jwst.nasa.gov/content/about/comparisonWebbVsHubble.html

## Chapter 7: High-energy Astronomy

Elizabeth Howell, 'Chandra Space Telescope: Revealing the Invisible Universe', Space.com, 16 June 2018, https://www.space.com/18669-chandra-x-ray-observatory.html

Bill Kendrick, 'About Uhuru', http://www.sonic.net/~nbs/projects/astro305-1/about/

'The Chandra X-ray Observatory', Harvard University, https://cxc.harvard.edu/cdo/about_chandra/overview_cxo.html

'Gamma Ray Bursts', NASA, https://imagine.gsfc.nasa.gov/science/objects/bursts1.html

'NASA Celebrates 25 Years of Breakthrough Gamma Ray Science', NASA, 7 April 2016, https://www.nasa.gov/feature/goddard/2016/nasa-celebrates-25-years-of-breakthrough-gamma-ray-science

'Technology for X-ray and Gamma-ray Detection', NASA, https://imagine.gsfc.nasa.gov/observatories/technology/

## Chapter 8: The Future

Brian Clegg, *Gravitational Waves: How Einstein's Spacetime Ripples Reveal the Secrets of the Universe* (London: Icon Books, 2018)

Andrew May, 'Giant Magellan Telescope', *How It Works*, issue 153, July 2021

'Euclid Overview', ESA, https://www.esa.int/Science_Exploration/Space_Science/Euclid_overview

'LISA Mission Summary', ESA, https://sci.esa.int/web/lisa/-/61367-mission-summary

'Roman Space Telescope: Frequently Asked Questions', NASA, https://roman.gsfc.nasa.gov/faq.html

# INDEX

**We inhabit a planet in peril. Our once temperate world is locked on course to become a hothouse entirely of our own making.**

*Hothouse Earth* provides a post-COP26 perspective on the climate emergency, acknowledging that it is now impossible to keep this side of the 1.5°C climate change guardrail. The upshot is that we can no longer dodge the arrival of disastrous – and all-pervasive – climate breakdown that will come as a hammer blow to global society and economy.

Bill McGuire explains the science behind the climate crisis and presents a blunt but authentic picture of the world bequeathed to our children and grandchildren; a world already glimpsed in today's blistering heatwaves, calamitous wildfires and ruinous floods and droughts. This picture is one we must all face up to, if only to spur genuine action to stop a harrowing future becoming a truly cataclysmic one.

**ISBN 978-178578-920-5 (paperback)**
**ISBN 978-178578-921-2 (eBook)**

**Music is shaped by the science of sound.**

How can music – an artform – have anything to do with science? Yet there are myriad ways in which the two are intertwined, from the design of instruments and hi-fi systems to how the brain processes music.

Science writer Andrew May traces the surprising connections between science and music, from the theory of sound waves to the way musicians use mathematical algorithms to create music.

The most obvious impact of science on music can be seen in the way technology has revolutionised how we create, record and listen to music. Technology has also provided new insights into the effects that certain music has on the brain, to the extent that algorithms can now predict our reactions with uncanny accuracy, which raises a worrying question: how long will it be before AI can create music on a par with humans?

**ISBN 978-178578-991-5 (paperback)**
**ISBN 978-178578-990-8 (eBook)**